Target Costing
and Value Engineering

STRATEGIES IN
CONFRONTATIONAL
COST MANAGEMENT
SERIES

TARGET COSTING
AND VALUE ENGINEERING

ROBIN COOPER

REGINE SLAGMULDER

PRODUCTIVITY PRESS
PORTLAND, OREGON

THE IMA FOUNDATION FOR APPLIED RESEARCH, INC.
MONTVALE, NEW JERSEY

Additional copies of this book and information about the Strategies in Confrontational Cost Management Series are available from the publisher. Discounts are available for multiple copies through the Sales Department (800-394-6868). Address all other inquiries to:

Productivity Press
P.O. Box 13390
Portland, OR 97213-0309
United States of America
Telephone: 503-235-0600; telefax: 503-235-0909
E-mail: service@ppress.com

Book and cover design by Bill Stanton, Productivity Press
Graphics and composition by Claire Barth, IMA/FAR
Printed and bound by Edwards Brothers in the United States of America

The IMA Foundation for Applied Research, Inc.
10 Paragon Drive
Montvale, NJ 07645-1760

Library of Congress Cataloging-in-Publication Data

Cooper, Robin, 1951–
    Target costing and value engineering / Robin Cooper, Regine Slagmulder
        p. cm—(Strategies in confrontational cost management series)
    Includes bibliographical references and index.
    ISBN 1-56327-172-9 (hardcover)
    1. Value analysis (Cost control) 2. Target costing.
I. Slagmulder, Regine. II. Title. III. Series.
658.5′75—dc21   97-15146
                                                                              CIP

IMA Publication Number 240-97319

05 04 03 02 01 00 99 98 97 10 9 8 7 6 5 4 3 2 1

This series is dedicated to the memory of

DAVID DENNIS GALLEY

This first volume is dedicated to our spouses,

HELEN and KENNETH,

for the many different ways in which
they have both supported us in this endeavor

CONTENTS

Foreword. xiii

Preface to the Series. xv
    Introduction   xv
    The Series   xvi
    The Firms   xix

Preface. xxi

Acknowledgments xxv

About the Authors. xxix
    Robin Cooper   xxix
    Regine Slagmulder   xxxi

About the Sponsors xxxiii
    Institute of Management Accountants   xxxiii
    Institute for the Study of U.S./Japan Relations
        in the World Economy   xxxiv
    Monitor Company, Inc.   xxxv

Figures, Tables, Exhibits xxxvii

Executive Summary 1
   Introduction   1
   The Confrontation Strategy   3
   The Survival Triplet and the Survival Zone   4
   Cost Management in a Confrontation Strategy   6
   Target Costing and Value Engineering   8
   Implications for Western Managers   16

PART ONE:

CONFRONTATIONAL COST MANAGEMENT

CHAPTER 1:

How Firms Compete Using the Confrontation Strategy 21
   Introduction   21
   Lean Production and Product Development   22
   The Three Generic Strategies of Competition   30
   Managing the Survival Triplet   36
   Surviving the Transition to Confrontational Competition   39
   Surviving in a Confrontation Mode   42
   Summary   46

CHAPTER 2:

The Role of Cost Management in Confrontation Strategy . . . 49
   Introduction   49
   Managing the Cost of Future Products   50
   Managing the Cost of Existing Products   53
   Harnessing the Entrepreneurial Spirit   57
   Summary   59

CHAPTER 3:

The Research Project 61
   Introduction   61
   The Research Approach   62
   Forming the Project Committee and Research Team   62
   Site Selection—The Series   62
   Site Selection—This Volume   64
   Visiting the Sites and Writing the Case Studies   66
   Analyzing and Synthesizing the Case Studies   67

PART TWO:

TARGET COSTING AND VALUE ENGINEERING

CHAPTER 4:

An Overview of Target Costing and Value Engineering 71
Introduction   71
Target Costing   71
Value Engineering   80
Summary   85

CHAPTER 5:

Market-Driven Costing 87
Introduction   87
Setting Long-Term Sales and Profit Objectives   89
Structuring the Product Lines   92
Setting the Target Selling Price   94
Setting the Target Profit Margin   100
Setting the Allowable Cost   104
Summary   106

CHAPTER 6:

Product-Level Target Costing 107
Introduction   107
Setting the Product-Level Target Cost   108
Disciplining the Product-Level Target Costing Process   119
Achieving the Target Cost   126
Summary   127

CHAPTER 7:

Value Engineering 129
Introduction   129
The Organizational Context   130
Planning the VE Process   132
VE Techniques   133
Summary   137

CHAPTER 8:

Component-Level Target Costing 139
Introduction   139
Setting the Target Costs of Major Functions   140
Setting the Target Costs of Components   150
Summary   163

CHAPTER 9:

Factors That Influence the Target Costing Process 165
Introduction   165
Factors That Influence Market-Driven Costing   167
Factors That Influence Product-Level Target Costing   174
Factor That Influences Component-Level Target Costing   181
How the Factors Influence the Target Costing Process   184
How Some of the Factors Influence Multiple Sections
    of the Target Costing Process   186
Comparing the Target Costing Processes   188
Summary   194

CHAPTER 10:

Target Costing and Value Engineering in Action 197
Introduction   197
The Acme Pencil Co. Ltd.   197
Market-Driven Costing   200
Product-Level Target Costing   208
Component-Level Target Costing   216
Summary   219

CHAPTER 11:

Lessons for Adopters 221
Introduction   221
Question One: Is profit management becoming more critical to your
    firm's survival?   222
Question Two: Is satisfying your customers becoming more critical to
    the survival of the firm?   223
Question Three: Is product design becoming more critical to your
    firm's survival?   225

Question Four: Are supplier relations becoming more critical to the
survival of your firm?   227
Question Five: Is cost management the right place for your firm to
expend resources?   228
Question Six: Can you create the right organizational context to
support target costing and value engineering programs?   230
Summary   234

PART THREE:

THE CASES

CHAPTER 12:

Nissan Motor Company, Ltd. 239

Introduction   239
Introducing New Products   241

CHAPTER 13:

Toyota Motor Corporation 253

Introduction   253
Globalization   254
Supplier Relationships   255
Cost Planning   256
Target Costing   257
Product Planning   257
Retail Prices and Sales Targets   258
Value Engineering   264
Knowing the Effect of Design Changes   265
From Cost Planning to Mass Production   266

CHAPTER 14:

Komatsu, Ltd. 269

Introduction   269
Construction Equipment   270
Product Development and Design for Manufacturability
    Cost Studies   270
Target Costing   274

CHAPTER 15:

Olympus Optical Company, Ltd.: Cost Management
   for Short Life-Cycle Products 281
   Introduction   281
   Strategic Change at Olympus   284
   Appendix   296

CHAPTER 16:

Sony Corporation: The Walkman Line 299
   Introduction   299
   The Walkman Product Line   300
   The Domestic Market   303
   The Overseas Market   306
   The Target Costing System   308
   New Product Development   309
   Product Costing   311

CHAPTER 17:

Topcon Corporation 313
   Introduction   313
   Ophthalmic Instruments   315
   Topcon's Production Control System   317

CHAPTER 18:

Isuzu Motors, Ltd. 327
   Introduction   327
   Cost-Creation Program   329
   Cost-Down Program   334
   Applying the Techniques   343

Appendix: Company Descriptions 347

Glossary of Terms 355

Selected Bibliography and References 361

Index 373

FOREWORD

Now, for the very first time, we are able to look behind the prover-
bial *shoji* screen and learn what only a few Westerners know—how
Japanese firms apply strategic cost management techniques. After
six years of field-based research with 25 firms, Dr. Robin Cooper
and Dr. Regine Slagmulder present to the outsider what really lies
behind Japan's global economic success. In a series of five volumes,
Robin, whose pioneering work in activity-based cost management
brought him world acclaim, and Regine reveal the cost management practices embedded in Japanese organizations.

Besides discovering what is being done, we are introduced to a
powerful new concept that has far-reaching strategic consequences
for global competitors: the *survival triplet*—cost-price (internal cost
structure/external selling price), quality (conformity to specifications), and functionality (what it does). We suddenly realize that
among lean enterprises, competition becomes confrontational. The sur-
vivor is the one who adopts an integrated, confrontational strategy

and becomes an expert at developing low-cost, high-quality products with the functionality customers demand. A firm must integrate quality, functionality, and cost management systems to survive. This theme is pervasive throughout the research.

Putting aside Japan's cultural differences, geography, politics, and the economics of accumulating global wealth, strategic cost management concepts emerge as cornerstones for global competition. The island nation has become a formidable global trading partner, competing unrelentingly to gain market share wherever the opportunity exists. For Westerners to survive, we need to recognize this competitive phenomenon and act accordingly.

The journey toward understanding Japanese cost management methods starts with Robin Cooper's *When Lean Enterprises Collide* (Harvard Business School Press), followed by this series of titles that incorporate the "nuts and bolts" of cases used at the Harvard Business School. These volumes give you a overall view of Japanese practices and their implications. Each book will give you a deeper knowledge of these global survival tools and help you learn how to apply the strategy in your company.

The Institute of Management Accountants, working with Productivity Press, is proud to bring the series to readers in the global community through its research affiliate, IMA's Foundation for Applied Research, Inc. (FAR). The Japanese successes can be replicated by others but not without radical change. The transformation is a must for global survival.

This research contributes significantly to the emerging body of knowledge about strategic cost management and its place in the broader subjects of management accounting and financial management for improved competitiveness. The series reflects the views of the researchers and not necessarily those of the IMA, IMA's Foundation for Applied Research, Inc., or the project committee.

Julian M. Freedman, CMA, CPA, CPIM
Senior Director
IMA's Foundation for Applied Research, Inc.

PREFACE TO THE SERIES

INTRODUCTION

Lean enterprises have faster reflexes than their mass production counterparts. They have the ability to design and launch products very rapidly. These fast reflexes render the sustainable competitive advantages that mass producers have relied on virtually impossible to achieve. Instead, lean enterprises compete by repeatedly creating temporary competitive advantages. Because they do not have sustainable competitive advantages, lean enterprises are forced to seek out competition. They adopt a generic strategy of confrontation, that is, they compete head on by trying to sell equivalent products to the same customers.

When firms compete in this manner, three product-related characteristics play a critical role in strategy formulation. These characteristics, known as the survival triplet, have internal and external forms. Internally they are the product's cost, quality, and functionality. Externally, they are its selling price, perceived quality, and perceived functionality. While the selling prices of products can be disconnected from their costs temporarily, if the firm is to remain profitable in the long run, costs must be brought into line with selling prices. The survival zone of a product identifies the range of values of the three characteristics that a product must have to be successful.

Firms that compete in a confrontational environment must develop and integrate their total quality management, product development, and cost management systems. The objective of this integration is to create a strategy based on developing products with the right level of functionality and quality at the right price. The key point is that firms must learn to view the process of managing the survival triplet as a total systems solution, not a collection of independent techniques.

While the way Japanese firms manage quality and time to market have been documented in depth elsewhere, the way in which Japanese firms manage costs has not attracted anywhere near as much attention and represents the missing piece of the puzzle of how lean enterprises compete. In response to this situation, the Institute of Management Accountants, Robin Cooper, and Regine Slagmulder agreed to carry out a joint research project based on studies of actual Japanese lean enterprises that have implemented advanced cost management programs. The primary objective of the project was to research, synthesize, and document key design issues and results, based on these companies' experiences.

THE SERIES

The purpose of the *Strategies in Confrontational Cost Management Series* is to begin the process of filling in the gap in the literature about Japanese cost management practices. The first volume in the series is *Target Costing and Value Engineering*. The seven cases in this volume document how firms use target costing and value engineering to design products that have the functionality and quality customers demand at a cost that allows the firms to make adequate profits when the products are sold at their target selling prices.

Target costing achieves its objectives by transmitting the competitive pressures faced by the firm to its product designers and suppliers. It creates a common language among the various functions involved in bringing products to market: production, product engineering, procurement, and marketing. It helps individuals in the various functions develop products that have the right functionality, quality, and cost. By decomposing the product-level target costs to the component level, firms can begin the process of transmitting the competitive pressures they face to their suppliers.

The second volume in the series, on chained target costing and other interorganizational techniques, contains cases documenting the way Japanese firms transfer cost management pressures across organizational boundaries. The pressure to become more efficient has caused many firms to try to increase the efficiency of supplier firms through interorganizational cost management systems. These systems have emerged because it is no longer sufficient to be the most efficient firm; it is necessary to be part of the most efficient supply chain. To achieve this objective, many Japanese firms blur their organizational boundaries. Organizational blurring occurs when information critical to one firm is possessed by another firm either further up or down the supply chain. The two or more firms then create relationships that share organizational resources, including information that helps improve the efficiency of the interfirm activities. Mechanisms for information sharing include supporting joint research and development projects, placing employees of one firm in other firms, and establishing interorganizational cost management systems.

These interorganizational cost management systems are designed to achieve three objectives:

- They create conduits that transmit the competitive pressures faced by the end producers to their suppliers.
- The systems let the product engineers at all firms in a supply chain jointly design products that can be manufactured more cost efficiently than if the firms acted independently.
- Through tradeoffs in the survival triplet, these cost management systems create a way to modify the specifications

that the end producer sets for the parts it purchases. These modifications allow the part to be sold at its target price while still generating an adequate return for all firms in the supply chain.

The third volume in the series explores kaizen costing, which is continuous improvement applied to cost reduction in the manufacturing stage of a product's life. The seven cases in the volume document the rich variety of practices surrounding kaizen costing. Unlike target costing, kaizen costing does not concern itself with changing a product's functionality. Instead, it keeps functionality constant and tries to find ways to reduce costs.

Kaizen costing has two major approaches: general and specific. General kaizen costing sets out to reduce the costs of production processes in general, and specific kaizen costing seeks to reduce the cost of a given product, either by changing its design or by reducing the cost of the production processes involved in its manufacture.

The next volume in the series is on product costing and operational control. The six cases in this volume capture a wide range of product costing and operational control practices. Product costing systems are used to report the cost of existing products so their profitability can be monitored. Reported product costs are used to help manage the mix of products offered, set selling prices, and identify products that require additional cost-reduction efforts, outsourcing, replacement, or discontinuance.

Operational control techniques are used to create cost-reduction pressures at the product and production process level. These techniques hold individuals responsible for the costs they can control and then report variances from expected results. The starting point for variances can be either standard costs or anticipated costs that reflect the savings from the firm's kaizen costing system.

The final volume in the series, on harnessing the entrepreneurial spirit, deals with cost management systems that achieve their objective in a different way than the other systems documented in the series. The six cases in this volume capture cost management techniques that do not focus on either the product or the production process but rather on modifying the behavior of the work force. Their effectiveness results from the increased pressure to perform placed on the leaders of the self-directed teams.

Two cost management techniques designed to achieve this effect were identified. The first technique is to convert cost centers into pseudo micro profit centers. The advantages associated with making cost centers into profit centers include increased responsibility and awareness of the effect of the groups' actions on the bottom line profitability of the firm. This increased awareness causes the group leaders to try to improve group performance in ways that increase profits. Since profits are simply revenues minus costs, this change in orientation leads to a more intense pressure to reduce costs and a new pressure to increase revenues. The second technique is to reduce effective firm size by breaking the firm into numerous, highly autonomous, real micro profit centers.

THE FIRMS

The material in this series was developed through observation and analysis of 25 innovative Japanese lean enterprises that have advanced cost management programs. We especially appreciate the cooperation from the companies we have worked with for allowing us to describe their cost management systems in depth. We were fortunate to receive permission to present the 31 cases, predominantly undisguised.[1] These firms were:

Citizen Watch Company, Ltd.
Higashimaru Shoyu Company, Ltd. (2) [2]
Isuzu Motors, Ltd. (2)
Jidosha Kiki Company
Kamakura Iron Works Company, Ltd.
Kirin Brewery Company, Ltd.

---

[1] Firms in italics are disguised.
[2] The number in parentheses identifies the number of cases written on the company.

Komatsu, Ltd. (2)
Kyocera Corporation
Mitsubishi Kasei Corporation
Miyota Company, Ltd.
Nippon Kayaku
Nissan Motor Company, Ltd.
Omachi Olympus Co., Ltd.
Olympus Optical Company, Ltd. (3)
Shionogi Pharmaceuticals
Sony Corporation
Sumitomo Electric Industries, Ltd.
Taiyo Kogyo Co., Ltd. (The Taiyo Group)
Tokyo Motors Company
Toyo Radiator Co., Ltd.
Toyota Motor Corporation
Topcon Corporation
Yamanouchi Pharmaceutical Co., Ltd.
Yamatake-Honeywell Company, Ltd.
Yokohama Corporation, Ltd.

In summary, this series represents a state-of-the-art analysis of the cost management systems at 25 Japanese manufacturing companies. The findings will provide insights into the nature of those systems, the conditions under which the techniques are likely to be most beneficial, and the detailed operations of such systems. Hopefully, these insights will be valuable in helping managers at firms that want to implement such systems achieve their objectives.

PREFACE

In today's intensely competitive environments, firms must become experts at developing low-cost, high-quality products that have the functionality customers demand. They must develop integrated quality, functionality, and cost management systems that ensure that products are successful when launched. These systems must create a firmwide discipline to design and manufacture products of high quality and functionality at low cost. Thus, the objective of the cost management program is to instill in everyone in the firm a disciplined approach to cost reduction. This discipline must begin when products or services are first conceived, continue during manufacturing, and end only when they are discontinued. The cost management program must not limit its scope to just the four walls of the factory or even the boundaries of the firm. It must spread across the entire supplier and customer chains.

Effective cost management must start at the design stage of a product's life. Once a product is designed, most of its costs are committed. For example, the number of components, the different types of materials used, the time it takes to assemble are all determined primarily by the way the product is designed. Some authorities estimate that as much as 90% to 95% of a product's costs are designed in; that is, they cannot be avoided without redesigning the product. Consequently, effective cost management programs must focus on the design as well as the manufacturing phase of a product's life cycle.

The primary cost management method used by many Japanese firms to control cost during the product design stage is a combination of *target costing* and *value engineering*. Target costing has two primary objectives. The first is to identify the cost at which the product *must* be manufactured if it is to earn its target profit margin at its expected or target selling price. The second objective of target costing is achieved by decomposing the target cost down to the component level. The firm's suppliers then are expected to find ways to deliver the components they sell at the target prices set by their customers while still making adequate returns. Value engineering is an organized effort to analyze the functions of goods and services and find ways to achieve those functions in a manner that allows the firm to meet its target costs. Like target costing, VE is applied during the design phase of product development and involves a multidisciplinary, team-based approach.

While target costing and value engineering have become cornerstones of Japanese cost management programs, they have not received as much attention in the West. Some Western firms have developed such systems, but many lack a fully integrated cost management approach to product design. The objective of this volume is to document and analyze how Japanese manufacturing firms use these techniques for strategic advantage. We hope that Western managers can adapt these techniques to the specific requirements of their firms.

The volume is structured as follows. An executive summary provides an overview of the findings of this portion of the research project. In Part One, the first chapter assists readers in understanding the terminology and concepts behind the confrontation strategy by describing confrontational principles and how to manage the survival triplet. Chapter 2 covers the role of cost management in a confrontation strategy. Chapter 3 describes the research method. Chapters 1 to 3 are common to all volumes in the series.

The seven chapters in Part Two concern the two main topics of this particular volume. Chapter 4 presents an overview of target costing and value engineering. It is followed by six chapters that summarize the outputs and process of target costing, based on an in-depth examination of the practices at seven companies. These insights are presented in Chapters 5 to 10. Chapter 11 discusses lessons for adopters.

The final part of the book contains seven case studies of Japanese manufacturing companies that have implemented target costing and value engineering systems. The case studies, documented in Chapters 12 to 18, highlight the competitive context that led management to implement target costing, the design of the firms' target costing systems, the application of target costing, and the value engineering process.

ACKNOWLEDGMENTS

This volume would not have been possible without extensive time commitment from a large number of persons. Some of these people were involved in the underlying research, others were responsible for editing the cases, and some were involved in reviewing the draft manuscripts. Many of them donated their time despite busy schedules.

First and foremost, we owe an enormous debt to individuals at the seven Japanese firms whose target costing and value engineering practices form the basis for the volume. It is impossible to say how many people in those firms, some of whom we never even met, provided input to the research, but the number is well over 50. To give some idea of the effort required on their part it should be noted that we spent a total of 20 days visiting the seven companies. When the cases were written, more than 500 outstanding questions had to be answered. At the end of the project, only a few remained unanswered, and none of them was considered important. Every case was read at least three times by persons at the companies, and corrections and suggestions were made after each reading. These suggestions helped to ensure the accuracy of the cases and the richness of our understanding of Japanese practice. We would particularly like to thank Toshiro Shimoyama, chairman and CEO of Olympus Optical Co., Ltd. for spending considerable time with us discussing the theory of confrontational strategy and the role of cost management.

We also owe a debt of gratitude to Professor Takeo Yoshikawa of Yokohama National University, who coauthored the case on Isuzu Motors, Ltd., and to Professor Takao Tanaka of Tokyo Keizai University, who wrote the original article from which the coauthored Toyota case was developed. Nissan and Komatsu were identified by Professor Michiharu Sakurai of Senshu University, to whom we express our thanks.

We would like to thank those who were actively involved in the process of writing the cases and this volume: Sarah Connor, Juliene Hunter, and Amy Wong. Each provided unflagging energy to keep the project on course. Sarah, a case editor at Harvard Business School case services, performed the invaluable and nearly endless task of editing the more than 700 pages of cases and teaching notes that make up the Japanese Cost Management Series. It was a delight to work with her. Amy and Juliene helped arrange the visits to the companies and assisted with editing the drafts of the analysis chapters of the volume. Without their help, the project would have taken much longer to complete.

Reviewers also offered support. They included Shannon Anderson, Marc Epstein, Jitsuo Goto, Robert Kaplan, Kentaro Koga, Charles Leinbach, Cecily Raiborn, and Takeo Yoshikawa. The following students enrolled in the MBA program at the Peter F. Drucker Graduate Management Center of The Claremont Graduate School contributed to the material contained in Chapter 11: Kimberly Green, Sandy Pilotnik, James Richardson, Mohan Sivasankar, and Moana Vercoe.

In addition we would like to thank all those at IMA and Productivity Press who helped shape this manuscript and bring it to bound book form: Mike Desposito of Productivity Press for establishing the partnership between Productivity and IMA; Julian Freedman, senior director of IMA's Foundation for Applied Research, Inc.; Diane Asay, editor in chief of Productivity; Claire Barth, senior editor of IMA, for doing both substantive and copy editing and typesetting, as well as acting as project manager; Gary Peurasaari of Productivity Press for editing suggestions; Susan Swanson of Productivity Press for managing the production process; and Bill Stanton, art director at Productivity Press, for page and cover design.

Clearly, this volume could not have come about without the help of institutions that supported the research. This was an extremely expensive undertaking, and we hope that all the institutions consider their money well spent. First, there are the academic institutions with which we have or have had associations and second, there are the sponsors of the series. For Robin Cooper, the first three years of the project were supported by the Harvard Business School, Division of Research. The last four years were supported by the Institute of U.S./Japan Relations in the World Economy and the Claremont Graduate School. Regine Slagmulder was supported by the Institute of U.S./Japan Relations in the World Economy and the University of Ghent.

The series has three sponsors. First, the Institute of Management Accountants had the foresight to see the need for the series. Their support and guidance contributed greatly to this project's success. The Project Committee of the IMA's Foundation for Applied Research, composed of Robert Miller (chairman) and Hank Davis, provided sustained enthusiasm and invaluable guidance as to the content and direction of each of the analysis chapters. In addition, The Monitor Company and the Institute of U.S./Japan Relations in the World Economy both provided financial support for the series.

About the Authors

Robin Cooper

Professor Robin Cooper has been a member of the faculty of the Peter F. Drucker Graduate Management Center, Claremont Graduate School, Claremont, California, since July 1992. Prior to that date he was on the Harvard Business School faculty from 1982 until 1992. He is also Visiting Professor of Strategic Cost Management at the Manchester Business School (U.K.). In 1996, he was awarded an honorary doctorate from the University of Ghent, Belgium.

His major field of interest is strategic cost management systems. His current research focuses on the design and use of cost management systems to achieve competitive advantage and, in particular, on Japanese cost management systems and activity-based cost systems.

In 1990, Cooper was the recipient of the first "Innovations in Accounting Education Award" presented by the American Accounting Association in recognition of his course development efforts in product costing. In 1991 and 1993, he was the recipient of the Notable Contributions to Management Accounting Literature presented by the Management Accounting Section of the American Accounting Association.

Cooper is a regular contributor to several journals including *Advances in Management Accounting, The Journal of Cost Management, International Journal of Production Economics, Management Accounting* (U.S.), *Management Accounting* (U.K.), *Accountancy* (U.K.), *Management Accounting Research* (U.K.), *Accounting Horizons, Accounting* (Japan), and *Sloan Management Review*. In addition, Cooper has had four articles published in the *Harvard Business Review*.

He is the author or coauthor of four books: *When Lean Enterprises Collide: Competing Under Confrontation*, published by the Harvard Business School Press; *Cost Management in a Confrontation Strategy: Lessons from Japan*, a customizable case book containing 22 cases on 19 Japanese firms; *Implementing Activity-Based Cost Management: Moving from Analysis to Action* (IMA, 1992), and *The Design of Cost Management Systems: Text, Cases, and Readings* (Prentice-Hall, 1991).

Cooper received his MBA with high distinction from Harvard in 1977 and was named a Baker Scholar. A recipient of a Deloitte Haskins & Sells Foundation Fellowship and an ITT International Fellowship, he earned his DBA from Harvard in 1982. Before beginning his graduate studies, he worked as an accountant for Coopers & Lybrand in its London and Boston offices from 1972 to 1976. He is a fellow of the Institute of Chartered Accountants in England and Wales. Cooper received his bachelor of science degree in chemistry with first-class honors from Manchester University in 1972.

REGINE SLAGMULDER

Professor Regine Slagmulder is a member of the faculty of De Vlerick School voor Management and the School of Applied Economics and Economic Sciences at the University of Ghent, Belgium. She is a research fellow of the Institute for the Study of U.S./Japan Relations in the World Economy at the Claremont Graduate School, Claremont, California. She is also a visiting professor of management accounting at Nijenrode University, The Netherlands.

Her current research focuses on strategic cost management systems and their applications in both Japanese and Western firms. She has had several articles published in international journals, including *Management Accounting Research* and *International Journal of Production Economics*. In addition, she is the coauthor of two books: *Management Accounting in de Nieuwe Productie-Omgeving* (Management Accounting in the New Production Environment) and *Management Control* (Beheerscontrole, Een Stimulans Voor Doelgericht Management van Organisaties).

Regine Slagmulder received her Ph.D. in management from the University of Ghent in 1995. Her doctoral research focused on the use of management control systems to align capital investment decisions with strategy. During her doctoral studies, she spent a year as visiting research fellow at Boston University School of Management. She earned a master's degree in electrical engineering in 1988 and a master's degree in industrial management in 1991, both from the University of Ghent.

About the Sponsors

Institute of
Management Accountants

The Institute of Management Accountants (IMA) is the world's largest organization devoted exclusively to management accounting and financial management. IMA has approximately 80,000 members and more than 300 chapters and affiliates across the United States and abroad. The IMA contributes to advancements in financial management and management accounting practices through education and professional certification. The IMA:

- publishes research reports and monographs on a wide variety of management accounting topics through its research affiliate, IMA's Foundation for Applied Research (FAR);

- provides continuing education courses and seminars to members;
- disseminates knowledge by publishing a monthly magazine, *Management Accounting*;
- awards the professional designations of Certified Management Accountant (CMA) and Certified in Financial Management (CFM).

Through such activities the IMA carries out its vision: *global leadership in education, certification, and practice of management accounting and finance.*

INSTITUTE FOR THE STUDY OF U.S./JAPAN RELATIONS IN THE WORLD ECONOMY

The Institute for the Study of U.S./Japan Relations in the World Economy is part of the Peter F. Drucker School of Management at Claremont Graduate University. Its mission is to advance Japan and U.S. economic, political, and social relations through intensive research and educational programs. To carry out its mission the Institute is developing partnerships globally with universities, research institutes, and sponsoring corporate foundations.

The Institute serves as an "objective and nonpartisan" resource center for the exchange of information and cross-fertilization of ideas for researchers and practitioners who are looking for solutions to reduce the economic, political, and social tensions between the United States and Japan. It also is an information exchange resource for the development of teaching and educational materials that can be used to train and educate current and future leaders in industry, academe, and government. It depends on an international network of multidisciplinary researchers (academics, research institutes, consultants, and government policy makers) to increase the influence of its findings on corporate leaders and public policy makers.

The Institute is unique in that it uses balanced research teams from the United States and Japan and strives to achieve balanced funding from U.S. and Japanese transnational corporations. By building a positive relationship with Japan through an innovative educational program that serves a global constituency, this program will help prepare current and future leaders for the strategic challenges ahead.

Claremont Graduate University in Claremont, California, is dedicated exclusively to graduate study, awarding degrees in 19 disciplines through six academic centers. A member of the Claremont Consortium, CGU combines relevant human-scale instruction with the facilities and academic breadth of a medium-sized university.

Monitor Company, Inc.

Monitor Company, Inc. was founded in 1983 and has emerged as a major presence in the consulting industry. It has grown from its original Cambridge base to 14 offices, with more than 600 professionals worldwide.

It was originally established to apply concepts being developed by members of the Harvard Business School's business strategy area, particularly the creation of competitive strategies built on sustainable competitive advantage, and has subsequently broadened its capabilities to include a wide variety of services. Mark Fuller, the company's CEO, taught at Harvard and co-founded the firm with Michael Porter who, as a professor on the Harvard faculty, continues as a member of the firm's board of directors and does all his consulting through Monitor.

Monitor has focused its consulting practice on working with clients to develop corporate strategies designed to bring change. The approach has been firmly grounded in strategic analysis and thinking (as well as implementation) and the integrated approaches required for success. The goal has been to work with the very senior levels of management, with the expressed objective of generating action that leads to improved competitiveness and profitability in complex and changing industry and corporate environments.

Activities in support of clients include:

- creating overall corporate strategies for maximizing shareholder value in highly diversified companies and those in transition, including assessments of the strategic competitiveness of individual business units and the role of corporate functions in the wealth creation of the company;

- developing competitive strategies for individual business units;
- designing strategies and tactics based on understanding customer groups and their needs;
- devising strategies for assessing and reshaping corporate portfolios, including efficient processes for divesting businesses that no longer fit the corporate objectives and strategies for entering new markets;
- designing and implementing logistics and distribution improvement programs;
- designing organization structures, management processes, and senior management reporting systems to support corporate and business unit strategies.

Within these broad fields, Monitor provides services that integrate traditional strategy domains with knowledge of human characteristics that motivate and inhibit action within organizations. Its investments in diagnosing the barriers to change in organizations have made Monitor a leader in areas such as competitive and industry simulation, economic modeling for complex systems, and the use of productive reasoning to overcome organizational inertia.

Monitor's commitment to integrating the behavioral and technical dimensions of addressing strategy has led to the belief that in most cases companies should maintain long-term relationships with their consultants, much like those maintained with key legal, financial, and other advisors. As a result, Monitor is structured around client relationships rather than industry practice areas.

Monitor has a large and diverse client base, which includes 100 of the Fortune 500 companies. Monitor's client relationship focus is reinforced by access to a wide network of world-class industry and technical specialists and academics to supplement its client knowledge and strategic expertise.

Figures, Tables, Exhibits

FIGURES

Figure 1. The Survival Triplet   5

Figure 2. The Survival Zone for a Product   5

Figure 3. The Target Costing Triangle   9

Figure 4. The Target Costing Process   11

Figure 1-1. Single-Piece Production   23

Figure 1-2. Batch-and-Queue Production   24

Figure 1-3. Lean Product Development   25

Figure 1-4. Major Function Design Teams   26

Figure 1-5. Managing the Design of Product Families   26

Figure 1-6. Conventional Batch-and-Queue
Product Development   27

Figure 1-7. The Survival Triplet   31

Figure 1-8. The Survival Triplet for a Product   33

Figure 1-9. The Survival Zone for a Product   34

Figure 1-10. The Survival Zones of the Cost Leader
and Differentiators   35

Figure 1-11. The Collapse of the Cost Leader and
Differentiator Survival Zones   35

Figure 1-12. A Confrontation Survival Zone   36

Figure 2-1. Managing the Cost of Future Products   50

Figure 2-2. Managing the Cost of Existing Products   53

Figure 2-3. Harnessing the Entrepreneurial Spirit   57

Figure 4-1. Committed Costs   73

Figure 4-2. The Target Costing Process   74

Figure 5-1. The Target Costing Triangle:
the Market-Driven Costing Section   88

Figure 5-2. Market-Driven Costing   89

Figure 5-3. Olympus Optical's Vertical and Horizontal
Differentiation Strategy   94

Figure 5-4. Setting the Target Selling Price   95

Figure 5-5. The Survival Zone of a Camera   98

Figure 5-6. Price and Functionality of a Camera over Time   99

Figure 5-7. Setting the Target Profit Margin   100

Figure 5-8. Nissan Motor Co. Ltd.: Identifying the
Target Profit Margin   101

Figure 5-9. Setting the Allowable Cost   105

Figure 5-10. Calculating the Allowable Cost   106

Figure 6-1. The Target Costing Triangle: The Product-Level
Target Costing Section   108

Figure 6-2. Product-Level Target Costing   109

Figure 6-3. Setting the Target Cost-Reduction Objective   110

Figure 6-4. The Strategic Cost-Reduction Challenge   110

Figure 6-5. The Target Cost-Reduction Objective   112

Figure 6-6. Determining the Strategic Cost-Reduction Challenge   112

Figure 6-7. The Cost-Reduction Path   115

Figure 6-8. Olympus Optical Company Ltd.: Introducing
the Product   118

Figure 6-9. Achieving the Target Cost   120

Figure 8-1. The Target Costing Triangle: Component-Level
Target Costing   140

Figure 8-2. Component-Level Target Costing   141

Figure 8-3. Setting the Target Costs of Major Functions — 142

Figure 8-4. The Toyota Matrix Structure — 143

Figure 8-5. Decomposing the Target Cost
to the Major Function Level — 144

Figure 8-6. Distributing the Target Cost
Across Major Functions — 146

Figure 8-7. Achieving the Target Cost — 147

Figure 8-8. The Reserve for the Production Manager — 149

Figure 8-9. The Target Costing Process Revised
for the Reserve for the Production Manager — 150

Figure 8-10. Setting the Target Costs of Components — 151

Figure 8-11. Decomposing the Target Costs
of Major Functions to the Component Level — 152

Figure 8-12. Setting the Target Cost of a Component — 153

Figure 8-13. Komatsu, Ltd.: Target Costing Via
Functional Analysis — 155

Figure 8-14. Komatsu, Ltd: Target Costing Via
Functional Analysis—Identifying the Target Cost — 156

Figure 8-15. Component-Level Target Costing — 163

Figure 9-1. The Interaction of Ease of Information
Collection and the Value of the Information — 167

Figure 9-2. Factors That Influence Market-Driven Costing:
Intensity of Competition — 168

Figure 9-3. Factors That Influence Market-Driven Costing:
Nature of the Customer — 170

Figure 9-4. Factors That Influence Product-Level Target Costing:
Product Strategy — 175

Figure 9-5. Factors That Influence Product-Level Target Costing:
Characteristics of the Product — 178

Figure 9-6. Factor That Influences Component-Level Target Costing — 182

Figure 9-7. How the Factors Influence the
Target Costing Process — 185

Figure 10-1. Acme Pencil: Profit Margin vs. Selling Price — 204

Figure 10-2. Acme Pencil: Market-Driven Costing — 205

Figure 10-3. The Target Cost-Reduction Objective — 211

Figure 10-4. Reducing the Cost of the New Eraser — 213

Figure 10-5. Acme Pencil: The Strategic
Cost-Reduction Challenge — 214

Figure 10-5. Acme Pencil: Monitoring Progress — 218

Figure 12-1. Identifying the Target Profit Margin   245
Figure 14-1. Cooling Capacity Versus Radiator Capacity   277
Figure 14-2. Radiator Capacity Versus Cost   277
Figure 18-1. Portion of Cost Deployment Flowchart.   331
Figure 18-2. Cost Strategy Map   333-334

TABLES

Table 4-1. Different Views of the Product   73
Table 4-2.The Component-Function Cost Matrix
   of a Pencil   84
Table 9-1. The Relative Values of the Influencing Factors   189
Table 10-1. Feature Weighting for Long-Life and
   Existing High-End Products   198
Table 10-2. The Selling Prices, Profit Margins, and Manufac-
   ing Costs per Dozen of the Existing Products   201
Table 10-3. Current Cost of Components for Long-Life and
   Existing High-End Products   209
Table 10-4. Current Manufacturing Costs of Long-Life
   and Existing High-End Products   210
Table 10-5. Current Costs of Long-Life and
   Existing High-End Products   210
Table 10-6. Current and As-If Cost of New
   Long-Life Product   211
Table 11-1. Supportive Direction for Each Factor   235
Table 15-1. Olympus Optical 1995 Financial Results   282
Table 17-1. Type and Frequency of Value-Added Projects   324

EXHIBITS

Exhibit 12-1. Life-Cycle Contribution Study   243
Exhibit 12-2. Breakdown of Product Profitability   252
Exhibit 14-1. Purpose of the A Study   271
Exhibit 14-2. Purpose of the B Study   272
Exhibit 14-3. Purpose of the C Study   272
Exhibit 14-4. Purpose of the D Study   273
Exhibit 18-1. Resume of Yoshihiko Sato   330
Exhibit 18-2. Typical Checklist   344

Target Costing
and Value Engineering

EXECUTIVE SUMMARY

INTRODUCTION

In today's highly competitive environment, firms must manage costs aggressively if they are to survive. Cost management must be applied across the entire life of the product by everyone in the firm. It must start at the earliest stages of a product's life because the ability to change the design of the product significantly increases the degree to which costs can be reduced. Products must be designed so that they deliver the quality and functionality that are demanded by customers while generating the desired level of profits for the firm.

One way to ensure that products are sufficiently profitable when launched is to design them to a target cost determined by subtracting the product's desired profit margin from its expected selling price. Under this approach cost is viewed as an input to the design process, not as an outcome of it. Forcing the product to achieve its target cost creates an intense cost discipline in the product design process. The cost management technique that is used to discipline the product design process in this manner is called target costing.

By setting target costs based on market-driven selling prices, target costing transmits the cost pressure that is placed on the firm by the marketplace to everyone involved in the product design process. Through this pressure, target costing focuses the creativity of the firm's product designers on developing products that satisfy customers and that can be manufactured at their target costs.

Cost management must extend beyond the four walls of the firm. By setting the target costs that establish the purchase price of externally acquired components, target costing transmits the cost pressure to the firm's suppliers. They, too, are forced to find new ways to reduce costs while simultaneously increasing functionality and quality.

Four questions determine whether a firm uses target costing.

- Early in the design process, does your firm identify the target cost of products by subtracting their desired profit margin from their expected selling price?
- Does your firm specifically design new products so that they can be manufactured at their target cost?
- Are product-level target costs achieved most of the time?
- Does your firm decompose the target costs of its products to the component level and use the resulting component-level target costs as the basis for negotiations with suppliers?

If the answer to any of the above questions is no, your firm is not taking full advantage of target costing. It risks introducing products that cost too much and are overdesigned relative to customer requirements.

Target costing differs from conventional cost management techniques by adopting a feed-forward perspective. The objective is to design costs out of products, not try to find ways to eliminate costs after the products enter production. In today's competitive environment, few firms can afford to ignore such a powerful mechanism to increase profits.

The Confrontation Strategy

Cost management and, in particular, target costing and value engineering have become more important in recent years because of the emergence of the lean enterprise. At the heart of the lean enterprise is the belief that single-piece flow is more efficient than batch-and-queue. The removal of all the queues and other inefficiencies associated with batch-and-queue systems enables lean enterprises to have faster reflexes, to enjoy economies of scale at lower production volumes, and to be inherently more efficient than their mass producer counterparts. They can produce products with higher quality and functionality, at lower cost, more quickly. These increased abilities are natural outcomes of the single-piece flow philosophy. Near-perfect quality is necessary because without the inventories inherent in batch-and-queue systems, the just-in-time production processes must stop when a defect is encountered until its cause is identified and corrected. In contrast, in a batch-and-queue system, the defective parts are put aside and attention switches to a different product. Therefore, high levels of defects can be tolerated.

Applying the same single-piece philosophy to product design enables firms to reduce significantly the time it takes to develop and launch new products. This objective is achieved by creating multifunctional teams, each responsible for the development of a single product, instead of splitting the product development process into a number of distinct steps, each managed by a different department. The result is an enhanced ability to compete on functionality. A firm that can develop new products every 18 months will compete on functionality in significantly different ways than one that takes a decade to develop a new product.

The emergence of the lean enterprise dramatically changed the way firms compete, but this fact was hidden by the slow rate of adoption of the new, single-piece manufacturing philosophy. This lag between the adoption of an innovation and a change in the competitive environment allowed managers to find other reasons for the intensified competition they faced. New theories of competition emerged to describe how to compete in the new environment, but these theories did not identify the change from a batch-and-queue to a single-piece philosophy as the primary driving force behind the new order.

The competitive environment of mass producers supports the generic strategies of *cost leadership* and *product differentiation*. Both these strategies are based on the assumption that a firm can develop and sustain competitive advantages and so avoid competition. By developing a sustainable cost advantage, the cost leader can offer products that are low in price and functionality. In essence, the cost leader avoids competition by saying, "Don't compete with me. If you do, I'll drop prices even lower and render you unprofitable."

In contrast, product differentiators develop a sustainable advantage in product development. Their products have higher functionality but sell at higher prices. They develop unique products or services that closely satisfy customers' requirements. In essence, they isolate a section of the main market and state, "This is my territory. I'm so good at what I do that attempting to compete with me is pointless."

With the emergence and spread of lean production, the competitive environment has undergone a slow, steady transformation from competition between mass producers to competition between lean producers. New theories of competition emerged during the late 1980s and early 1990s. Competition between lean enterprises is based on the assumption that sustainable, product-related competitive advantages are unlikely to be developed. Since in the eyes of these firms there is no mechanism to avoid competition, they confront it and compete head on. Confrontation is necessary because lean enterprises can react fast enough to make product-related competitive advantages too fleeting to be considered sustainable. There is not enough time to educate the customer to the positive attributes of the new product before other firms have me-too versions. Unlike occasional head-on competition between mass producers, head-on competition between lean producers, once they have become confrontational, is continuous. This difference in the attitude of lean producers toward competition makes obsolete many of the lessons learned about competition between mass producers.

THE SURVIVAL TRIPLET
AND THE SURVIVAL ZONE

To understand when it is appropriate to adopt each of the three generic strategies—cost leadership, differentiation, and confronta-

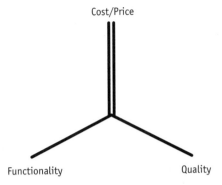

FIGURE 1. THE SURVIVAL TRIPLET

tion—it is helpful to introduce the concept of the *survival triplet* (see Figure 1). The survival triplet consists of the three dimensions that define a product, which are *cost/price, quality,* and *functionality.*

Only products with values along each of these three dimensions that are acceptable to the customer stand a chance of being successful. To identify a product's *survival zone* the *survival range* for each characteristic in the survival triplet must be determined. The survival range is defined by determining the minimum and maximum values that each characteristic can have for a product to be successful. The survival zone is the volume created by connecting the three minimum values and the three maximum values (see Figure 2).

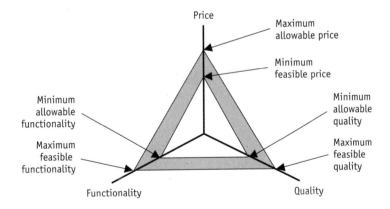

FIGURE 2. THE SURVIVAL ZONE FOR A PRODUCT

Cost leadership and differentiation strategies are successful when the survival zones for a firm's products are large. Large survival zones occur when the difference between the minimum and maximum ranges is significant for at least two of the characteristics. As the gap between the minimum and maximum levels widens, the ability of firms to create distinguishable products that have high values on one characteristic and low values on the others increases. When the gap becomes sufficiently large, it becomes possible to split the zone into at least two new survival zones, one based on low price and the other on enhanced functionality and quality. When this situation emerges, firms must choose to compete on either the price characteristic or on the other two. Firms competing on the price characteristic are vying for the cost leadership position, while firms competing on functionality and quality characteristics have adopted differentiation strategies.

When lean enterprises compete, survival zones become very narrow and becoming a differentiator is no longer possible. The bulk of customers in the industry are unwilling to make significant trade-offs among the three dimensions: there is simply not enough leeway for a firm to be able to differentiate its products and sell them at a sufficient price premium to justify the increased costs. All firms offer products that have similar high quality and functionality at a low price, which requires them to be produced at a low cost. Thus, confrontation strategies apply when, in effect, multiple firms compete for the same customers by developing equivalent products. Since the firms are evenly matched on all three dimensions, they have no choice but to compete head on and adopt a confrontational strategy.

COST MANAGEMENT
IN A CONFRONTATION STRATEGY

With the emergence of the lean enterprise and global competition, firms face ever-increasing levels of competition. As competition becomes more intense, firms are forced to learn to be more proactive in the way they manage costs. For many of these firms, survival depends on their ability to develop sophisticated cost management systems that create intense pressures to reduce costs over the entire life of the product and across the entire value chain. This increased importance of cost management is a central theme of *When Lean Enterprises Collide:*

> Firms that adopt a confrontation strategy must become experts at developing low cost, high quality products that have the functionality customers demand....A firm that fails to reduce costs as rapidly as its competitors will find its profit margins squeezed and its existence threatened....Cost management, like quality, has to become a discipline practiced by virtually every person in the firm. Therefore, overlapping systems that create intense downward pressures on all elements of costs are required.[1]

These systems manage costs in three ways. The first way is to manage the cost of future products, the second is to manage the cost of existing products, and the third is to harness the entrepreneurial spirit of the workforce. Japanese firms have developed three specific cost management techniques to manage the costs of future products: *target costing, value engineering,* and *interorganizational cost management systems.* Target costing is a structured approach to determine the cost at which a proposed product with specified functionality and quality must be produced to generate the desired level of profitability over its life cycle when sold at its anticipated selling price. Value engineering is used in the product design stage to find ways to achieve the specified functionality at the required standards of quality and reliability and at the target cost. To achieve the target cost without making sacrifices in product functionality and quality, interorganizational cost management systems are designed to create downward cost pressures on the entire supplier chain. The objective of these interorganizational systems is to identify innovative ways to reduce the cost of the components supplied by the chain.

Three cost management techniques are used to help manage the costs of existing products: *kaizen costing, product costing,* and *operational control.* Kaizen costing systems focus on making improvements to the production process of existing products. These

---

[1] Robin Cooper, *When Lean Enterprises Collide: Competing Through Confrontation,* Boston: Harvard Business School Press, 1995, p. 7.

improvements are designed either to increase the effectiveness of the production process in general or to reduce the costs of a specific product without altering its functionality.

Product costing systems are used to report the cost of existing products so that their profitability can be monitored. Using reported product costs, the firm can begin the process of identifying products that require redesign or discontinuance or that should be the focus of a specific kaizen program.

Operational control systems are used to monitor performance on the shop floor. They include techniques such as identifying responsibility centers, calculating variances, and providing feedback on performance.

A firm can lower the costs of its products in at least one other way: by harnessing the entrepreneurial spirit of its workforce. This method focuses on the workforce, not the products or production processes. Two techniques are used to accomplish this objective. The first creates *pseudo micro profit centers* from cost centers, and the second converts the firm into multiple *real micro profit centers*.

The objective of the cost management programs in many Japanese manufacturing firms is to create continuous pressure for cost reduction over the entire life of the product and across the entire value chain. This pressure must begin when products or services are first conceived, continue during manufacturing, and end only when the product or service is discontinued. Target costing and value engineering are the two major techniques used within the firm to manage costs during the product design stage.

TARGET COSTING AND VALUE ENGINEERING

Target costing is primarily a technique for profit management (see Figure 3). Its objective is to ensure that future products generate the profits identified in the firm's long-term profit plan. This objective can be achieved only if products satisfy the demands of the firm's customers and can be manufactured at their target costs. Transmitting the competitive pressure that the firm faces in the marketplace to the product designers and suppliers via *market-driven costing* promotes the necessary level of aggressive cost management. Subtracting the target profit margin from the target selling price set by the market identifies the cost at which the product must be manufactured. *Life-cycle costing* ensures that the target profit margins take into account the up-front investment in product development and any cost savings that can be anticipated over the product's life.

To create intense pressure on the product designers to reduce costs, *product-level target costing* focuses designer creativity on reducing the costs of future products to their target levels. These targets are set so that they can be achieved only if the product designers expend considerable effort designing costs out of future products. *Value engineering* is the primary technique used to find ways to decrease product costs while maintaining the functionality and quality the customer demands. It is key to achieving the target cost. As such it is an integral part of target costing.

To create an equivalent pressure on their suppliers, firms use *component-level target costing* to focus supplier creativity on reducing the costs of the components they supply. *Interorganizational costing* creates relationships that link customers and suppliers with

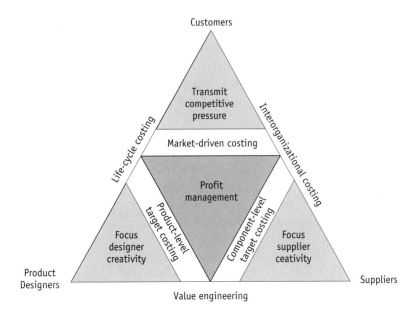

FIGURE 3. THE TARGET COSTING TRIANGLE

the firm's design engineers, helping them find ways to design lower-cost products.

Firms use target costing to ensure that new products are profitable when launched. Target costing is *a structured approach to determine the life-cycle cost at which a proposed product with specified functionality and quality must be produced to generate the desired level of profitability over its life cycle when sold at its anticipated selling price.*[2] Discussing target costing in terms of the survival triplet is valuable because the latter explicitly links product price and cost with quality and functionality. Since target costing relies on target selling prices it is necessary to take customers' quality and functionality requirements into consideration when setting target costs. Without this link, there are no constraints on the target costing process, and it is possible to set too low a target cost to enable the product to be manufactured with acceptable levels of quality and functionality. To be effective, target costing must take into account the constraints of product quality and functionality.

Target costing systems rely heavily on the cardinal rule: *"The target cost of a product can never be exceeded."* Without this rule, target costing systems typically lose their effectiveness. The primary objective of the cardinal rule is to stop the steady and relentless creep in product functionality and cost that occurs when product designers say, "If we just add this feature, the product will be so much better (and cost only a little more)."

The target costing process can be broken into three major sections. The first section identifies the allowable cost of each product. This is the cost at which the product must be manufactured if it is to earn its target profit margin at its expected target selling price. The second section identifies the product-level target cost, which is set to be achievable, but only if the product designers expend considerable effort and creativity. The third section identifies the component-level target costs. The firm's suppliers are then expected to find ways to deliver the components at their target costs while still making adequate returns themselves (see Figure 4).

---

[2] Target costs should include any costs that are driven by the number of units sold. For example, if the firm accepts responsibility for disposing of a product at the end of its useful life, these costs would be included in the target cost.

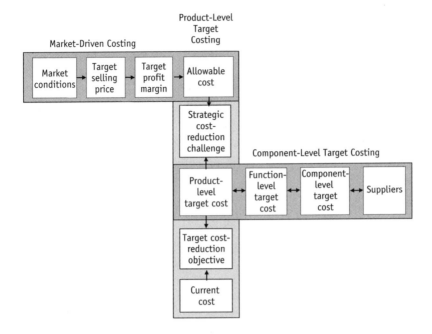

FIGURE 4. THE TARGET COSTING PROCESS

Market-Driven Costing—
Setting Allowable Costs for Products

Target costing begins with an analysis of the market conditions. The aim is to determine the selling price of a future product before it has even been designed. The organization must specify the quality and functionality of the future product and demonstrate that the product will be successful when launched at its target selling price.

The typical starting point for determining the target profit margin for a new product is the historical margin such products have earned. If all else remains the same, this is the margin that will be used to determine target costs. However, the relative strength of competitive offerings can cause the historical margin to be modified. Many firms use computer simulations to help them estimate overall firm profitability based on individual target profit margins.

Target profit margins have to reflect the economics of the product's life cycle. As a result, companies that are creating products with long development cycles and significant investment in product development often use life-cycle target costing methods that allow for the high capital expenditure associated with product development. Similarly, for firms that can substantially reduce the cost of their products during the manufacturing stage of the product's life cycle, the effect of potential cost reductions made by both the firm and its suppliers is included in the financial analysis of the product's profitability. Once the appropriate price, functionality, and quality targets have been set, then the *allowable cost* can be determined by subtracting the target profit margin from the target selling price.

Product-Level Target Costing

The allowable cost does not take into account the cost-reduction capabilities of the firm's product designers or suppliers. When the allowable cost of a product is considered unachievable, a higher *product-level target cost* is established. This cost is derived by starting with the *current cost* (the cost at which it could be manufactured if launched today) of the future product and subtracting the *target cost-reduction objective*. This objective is set so that it is achievable but only with considerable effort and creativity on the part of the product designers. The difference between the target cost and the allowable cost is the *strategic cost-reduction challenge.* This challenge captures the inability of the firm to achieve its long-term profit objectives for the product.

To be effective, target costing systems must create and maintain an intense discipline to reduce costs during the product design stage. This discipline is achieved in two ways—by continuously monitoring the progress the product design teams are making toward their cost-reduction objectives and by applying the cardinal rule of costing, that the target cost can never be violated. The cardinal rule is applied by:

- ensuring that whenever target costs increase somewhere in the product, costs are reduced elsewhere by an equivalent amount,
- refusing to launch products that are above their target costs,
- managing the transition to manufacturing to ensure that the target cost is indeed achieved.

Like all rules, the cardinal rule is sometimes violated. Violations are allowed when launching the product is more beneficial than not doing so. There are several reasons for this condition, most of which have to do with lost sales of other products either now or in the future.

Once the product-level target cost has been set, the designers have to find ways to design the product so it can be manufactured for that cost. The primary technique for achieving this objective is value engineering.

Value Engineering

Simultaneously with the target costing process, the firm uses value engineering to find ways to increase functionality and quality while meeting target costs. VE thus forms an integral part of target costing. It is a systematic, interdisciplinary examination of factors affecting the cost of a product so as to find means to fulfill the product's specified purpose at the required standards of quality and reliability and at an acceptable cost. It accomplishes this objective by analyzing products to find ways to achieve their necessary functions and essential characteristics. The necessary functions define what the product must be able to do. For example, a pencil must be able to mark the paper. The essential characteristics are the other requirements that must be satisfied if the product is to be successful, such as reliability, maintainability, and quality. VE helps to manage the trade-off between functionality and cost, the two dominant characteristics of the survival triplet. The objective of most Japanese VE programs is not to minimize the cost of products, but to achieve a specified level of cost reduction established by the firm's target costing systems.

There are four major ways to apply VE principles to product design. These ways are zero-look, first-look, and second-look value engineering, and teardown. Zero-look VE is the application of VE principles at the concept proposal stage, the earliest stage in the design process. Its objective is to introduce new forms of functionality that did not previously exist. First-look VE focuses on the major elements of the product design and is defined as developing new products from concepts. The objective is to enhance the functionality of the product by improving the capability of the existing functions.

Second-look VE is applied during the last half of the planning stage and the first half of the development and product preparation stage. The objective of second-look VE, unlike that of zero- and first-look VE, is to improve the value and functionality of existing components, not create new ones. Consequently, the scale of changes is much smaller than for zero- and first-look VE. The objective of teardown is to analyze competitive products in terms of the materials they contain, the parts they use, and the way they function and are manufactured.

Component-Level Target Costing

Once the target cost of a new product is established, multifunctional product design teams decompose it to determine component-level target costs. The cost-reduction objective is allocated across the components and subassemblies that make up the product, but it is not spread evenly across all of them. Historical trends, competitive designs, and other data are used to estimate how much cost can be removed from each component or subassembly. An iterative process ensures that when the component and subassembly target costs are set, they add up to the product's target cost.

At many firms, the chief engineer is responsible for establishing the main themes of the new product. For example, for an automobile it might be a quieter but sportier ride, while for a camera it might be the zoom range of the lens assembly. To achieve the main themes, the chief engineer may allocate more costs to those features, but under the cardinal rule every extra dollar allocated to improving a feature of the product must be taken out somewhere else.

The target costs of the components in most cases become the selling prices of the firm's suppliers. Thus, target costing transmits the competitive reality faced by the firm to its suppliers. The only exceptions are when the firm lacks the power to enforce selling prices on its suppliers or when suppliers can demonstrate that the target costs are unrealistic.

Factors Influencing the Target Costing Process

At least five major factors influence the target costing process. Two of them influence primarily the market-driven costing portion. They are the intensity of competition and the nature of the customer. The next two factors influence the product-level target costing process. They are the firm's product strategy and the characteristics of the product. Finally, the last factor, the firm's supplier-base strategy, shapes component-level target costing.

The *intensity of competition* influences how much attention the firm should pay to competitive offerings because they help shape customer requirements. Three aspects of the *nature of the customer* help determine the benefits derived from target costing. They are the customer's degree of sophistication, the rate at which future customer requirements are changing, and the degree to which customers understand their future product requirements.

Market conditions not only affect the benefits that can be derived from the application of target costing, they also determine the appropriate *product strategy* the firm should adopt. The product strategy is a primary determinant of the number of products the firm launches, the frequency of redesign, and the degree of innovation. Three aspects of the *product characteristics* have a particularly strong influence on the benefits the firm derives from target costing and the way it is practiced. These characteristics are the complexity of the product, the magnitude of the up-front investment, and the duration of the product-development process.

The *supplier-base strategy* shapes the firm's interactions with its suppliers. In particular, it determines the degree of horizontal integration, the firm's power over its suppliers, and the nature of supplier relations.

It is not possible to discuss the effects of these five factors in isolation. Rather, it is the subtle interaction of all of them that determines whether target costing will be beneficial, how beneficial it will be, the relative importance of the three parts of the target costing process, and the intensity and formality of the process.

Lessons for Adopters

Potential adopters should ask themselves six fundamental questions before beginning to implement target costing and value engineering. These questions help identify the likely benefits from the adoption of the two feed-forward techniques. The first, is profit management becoming more critical to the survival of your firm?, explores whether the firm is facing increasingly intense competition. Underlying this question is the premise that as competition becomes more intense, cost management becomes more important to the success of the firm.

The second to fourth questions focus on the three major sections of target costing. The first of these questions, is satisfying your customers becoming more critical to the survival of the firm?, focuses on customer requirements and the need to shift to price-driven costing. The next question, is product design becoming more critical to your firm's survival?, explores the way changes in the firm's product strategy and product characteristics are altering the benefits derived from product-level target costing. The third question of this group, are supplier relations becoming more critical to the survival of your firm?, looks at the changing role of suppliers and the effect of such changes on the benefits derived from component-level target costing.

The fifth question, is cost management the right place to expend resources?, checks whether the firm is facing sufficient competitive pressure to ensure that the benefits warrant implementing sophisticated target costing and value engineering systems. As profit margins fall, the relative benefits of undertaking cost management increase compared to trying to increase sales.

The final question, can you create the right organizational context to support target costing and value engineering programs?, explores management's ability to create an environment that supports the autonomous, multifunctional teams that lie at the heart of target costing and value engineering. Such an environment includes the appropriate reward structures, conditional autonomy for the teams, and the existence of champions for both the analysis and action phases of the target costing and value engineering implementation project.

IMPLICATIONS FOR WESTERN MANAGERS

Under the confrontation strategy, it is not necessary or advisable to expend equal effort on all three characteristics of the survival triplet. One characteristic usually dominates the other two. The most important of the three characteristics of the survival triplet to the firm's customers, and hence to the firm, frequently changes over time. When the Japanese economy went into severe recession in the early 1990s, many Japanese firms changed their most important characteristic from functionality to cost.

The key to success lies in selecting the appropriate rate of improvement for each characteristic. In a market where the customer is demanding increased functionality and is willing to pay for it, for example, the most important characteristic is functionality. The firm that can increase the functionality of its products fastest (subject to cost-price and quality constraints) will develop a temporary competitive advantage.

Unfortunately, many Western managers have failed to understand the role of the survival triplet in confrontation strategy and have adopted a rallying call of highest quality, lowest cost, and first-to-market products. No firm can reasonably expect to be number one in all three elements of the survival triplet. Any firm that actually achieved this distinction for any length of time would dominate its competitors. Indeed, if it could sustain this advantage, it would become a monopoly because all its competitors would be bankrupt. Western firms have adopted this "best in all three" approach because they have encountered Japanese competitors that are superior to them on all three counts. To survive, these firms have had to improve simultaneously on all three characteristics. The resulting struggle for survival has caused many Western managers to lose sight of the critical fact that in most markets one element of the triplet is more important than the others. Once a product is inside its survival zone, the firm no longer has to improve its performance along all three elements equally aggressively. Instead, it has to learn to compete intelligently and to select the rate at which it improves performance on each element of the survival triplet.

Western managers who do not adjust their mode of competition will risk their firms. Firms that cling to the concept of sustainable competitive advantages and invest resources only in businesses that they believe have sustainable advantages will discover investment opportunities decreasing over time. Firms that cling to the traditional strategies of cost leadership and differentiation will discover that their ability to maintain these strategies is eroding. They will be forced to retreat from their markets as lean competitors outmaneuver them. Niche players may find themselves maneuvered into low-growth niches, as lean competitors launch products directly at them.

A similar fate awaits firms that try to cling to their historical profit margins. Competition between lean enterprises is fiercer than competition between mass producers, and overall profit margins are smaller. Retreating from products that have lower profit margins will be successful only if other, higher-margin products can be identified. However, in a confrontational environment high profits typically signify products that are in small volume or dead-end niches. Therefore, there is a significant risk that chasing high-margin products will lead to a long-term, unacceptable strategic position.

To survive the low margins, firms that adopt a confrontation strategy must become experts at developing low-cost, high-quality products with the functionality customers demand. The cost, quality, and functionality expertise required for this result must be used to form a coherent strategy based on developing products with the right level of functionality and quality at the right price. Consequently, firms adopting a confrontation strategy must develop integrated quality, functionality, and cost management systems.

It is the integration of these systems that allows many Japanese firms to respond so quickly to changes in economic conditions and to match the innovative products of their competitors. If these systems were stand-alone systems, a fast response rate would not be possible. Unfortunately, in Western literature the systems that have evolved to manage quality (typically described as "total quality management systems") and functionality (typically described as "time to market systems") have been described in isolation, and the systems that have emerged in Japanese firms to manage costs have been almost ignored. In a world of nonsustainable competitive advantage, however, costs have to be managed both aggressively and intelligently. A firm that fails to reduce costs as rapidly as its competitors will find its profit margins squeezed and its existence threatened. It is no longer good enough to say, "Reduce costs by 10 percent across the board." Cost management, like quality, has to become a discipline practiced by virtually every person in the firm. Therefore, integrated systems that create intense downward pressures on all elements of costs are required.

PART 1

CONFRONTATIONAL COST MANAGEMENT

CHAPTER 1

How Firms Compete Using the Confrontation Strategy

Introduction

Manufacturing in the 20th century has experienced two major revolutions. The first was the development of *mass production*, best exemplified by Henry Ford's Model T, and the second was the development of *lean production* by Toyota. Although it took both innovations only 10 years to evolve and mature—mass production from approximately 1915 to 1925 and lean production from 1951 to 1961—their spread to other countries and industries occurred more slowly. Indeed, it is still possible to identify firms that are lean but have yet to adapt to all the implications of being lean.

The slow spread of these innovations made subsequent evolution in the competitive environment appear independent of the changes in production philosophy. Firms that adopted one or the other of the innovations saw the nature of competition shift only as other firms in the industry changed their production philosophy. That is, lean competition occurs only when a sufficient percentage of the entire industry has converted to lean production. Until that time, the lean producers compete as if they were mass producers and thus make higher profits. This lag between the adoption of an innovation and any associated changes in the competitive environment allowed managers to find other reasons for the intensified competition they faced. New theories of competition emerged in the late 1980s and early 1990s to describe how to compete in the new environment, but these theories did not identify the change in production philosophy as the primary driving force behind the new order. Thus, the slow spread of these innovations obscured the result of their adoption—change in the nature of competition. Mass producers compete differently from craft producers; lean producers compete differently from mass producers.

The competitive environment of mass producers supports the generic strategies of cost leadership and product differentiation.[1] Both these strategies are based on the assumption that a firm can develop and sustain competitive advantages and therefore can avoid competition.[2] In theory, in their purest forms cost leadership and product differentiation strategies create zones of no competition. Consequently, it is unusual to see the best firms in an industry dominated by mass producers engaged in head-on competition.

With the emergence and spread of lean production, the competitive environment has undergone a slow, steady transformation from competition between mass producers to competition between lean producers. Consequently, before managers can determine how their firms should evolve over the next decade, they need to understand that lean enterprises do not compete in the same ways as mass producers.

LEAN PRODUCTION AND PRODUCT DEVELOPMENT[3]

At the heart of the lean enterprises is the central premise that single-piece is more efficient than batch-and-queue processing. In single-piece flow systems, an individual or a group manufactures or de-

---

[1] Michael E. Porter, *Competitive Strategy: Techniques for Analyzing Industries and Competitors,* New York: The Free Press, 1980.
[2] Pankaj Ghemawat, "Sustainable Advantage," *Harvard Business Review,* September–October 1986, pp. 53-58.
[3] This section draws heavily from James P. Womack and D. T. Jones, *Lean Thinking,* New York: Simon & Schuster, 1996.

signs only one product at a time. In addition, the individual or group is typically responsible for the entire process. In the single-piece approach, there are no meaningful queues between each operation—the next operation in the sequence "pulls" the product through the previous operation (see Figure 1-1). Thus, as soon as the previous operation is completed, the next operation is by definition ready to receive it.

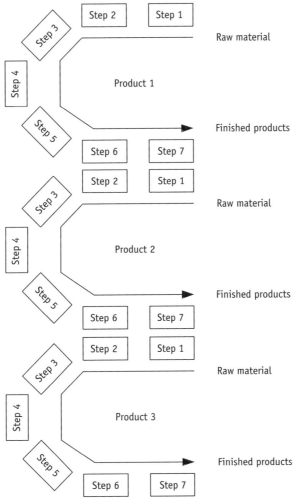

Adapted with the permission of Simon & Schuster from LEAN THINKING by James P. Womack and Daniel T. Jones. Copyright © 1996 by James Womack and Daniel Jones.

FIGURE 1-1. SINGLE-PIECE PRODUCTION

In contrast, in a batch-and-queue process products are produced or designed in batches. Each individual or group is responsible for only a part of the overall process. When that part is completed, the batch is sent to the next operation, where it waits in a queue until that individual or group is ready to start their portion of the overall process on the batch (see Figure 1-2). The product is pushed, not pulled,

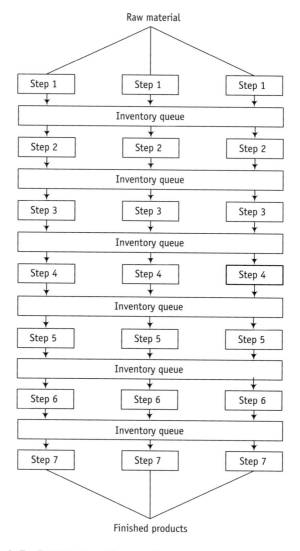

FIGURE 1-2. BATCH-AND-QUEUE PRODUCTION

through the process. Hence, there is no guarantee that the next operation in the process will be ready, and extensive queues typically result. The firms in the sample have applied single-piece principles wherever possible throughout the value chain of the product. They all use lean production philosophies, producing products one at a time. They also have adopted lean product development philosophies. At the heart of lean product development is an individual responsible for the entire development of the new product and sometimes for the product over its entire life. This directly responsible person, usually called the chief engineer, is supported by a multifunctional design team. The design team frequently includes people from marketing, engineering (often several branches), purchasing, production, and major suppliers. These teams are co-located and frequently have no other responsibilities than developing the product for which they were created (see Figure 1-3). The objective of creating these design teams is to enable the product development process to flow as rapidly as possible. Simultaneous engineering also is frequently used to speed up the product development process.

As the complexity of the product increases, it becomes beneficial to break the product into its major functions and create teams that are each responsible for one major function (see Figure 1-4). These

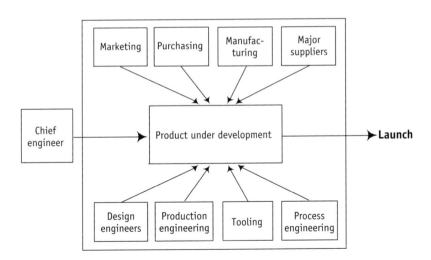

FIGURE 1-3. LEAN PRODUCT DEVELOPMENT

	Design team product A	Design team product B	Design team product C	• • •	Design team product M
Major function 1 division	Design team A1	Design team B1	Design team C1		Design team M1
Major function 2 division	Design team A2	Design team B2	Design team C2		Design team M2
Major function 3 division	Design team A3	Design team B3	Design team C3		Design team M3
				• • •	
Major function n division	Design team An	Design team Bn	Design team Cn		Design team Mn

FIGURE 1-4. MAJOR FUNCTION DESIGN TEAMS

teams are also co-located and again have no other responsibilities. The major function design teams report to the product design team and in some of the firms also to the head of the division responsible for designing that major function for all the firm's products. Thus, the entire product design process is single-piece with each team dedicated to the entire product or a major function of it.

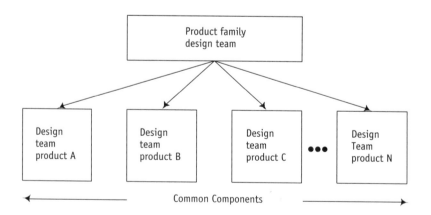

FIGURE 1-5. MANAGING THE DESIGN OF PRODUCT FAMILIES

In firms with complex product mixes, it is often helpful to identify product families. Each family consists of a number of similar products that can be produced interchangeably in a single production cell. It is useful to identify such families at the design stage, since they can usually be designed to share a large percentage of common components. To achieve this objective, product design is often managed at the product-family level. Each product family has its own design team led by a senior chief engineer who provides common guidance to the chief engineers responsible for the individual products in the family (see Figure 1-5).

Lean product development can be contrasted to conventional product development, in which the development process is not the responsibility of a single team but is distributed across multiple departments (see Figure 1-6). This departmentalized approach leads to batch-and-queue product development. The product definition is developed by the marketing department and transferred

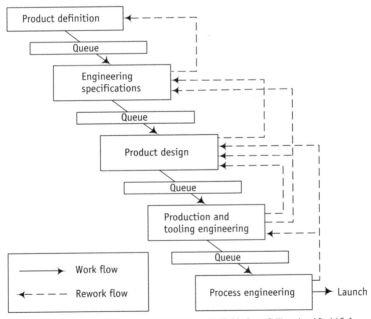

Adapted with the permission of Simon & Schuster from LEAN THINKING by James P. Womack and Daniel T. Jones. Copyright © 1996 by James Womack and Daniel Jones.

FIGURE 1-6.   CONVENTIONAL BATCH-AND-QUEUE
PRODUCT DEVELOPMENT

to engineering, where it is converted into engineering specifications. These specifications are then converted into blueprints and, when the design is complete, handed over to production and tooling engineering. When these departments have completed their work, the final design is handed to process engineering for final review.

On the surface, this process sounds highly efficient. However, it hides a multitude of sins. First, the specialization and location of skills means that at each new stage, the design often has to be reworked to make it more feasible.

Second, the design typically undergoes a steady transformation during the product development process as the market-based knowledge provided by marketing becomes more and more remote. This transformation occurs in two ways: through redesign requests to departments earlier in the process and through "secret redesign of earlier stages" by departments later in the process to make the product design feasible. These redesigns can lead to products that do not satisfy the customer because they lack critical functionality, have excess functionality, or cost too much to be sold profitably at an acceptable price.

Finally, each department may have a backlog of work so that there might be considerable queuing time between each step. This queuing time can add considerable delays to the product development process and lead to an inability to launch new products on a timely basis.

In contrast, lean product development streamlines the entire product design process by removing all the redesign steps and the queues between departments. The single-piece design team works much like a lean production cell. All the members can see the entire design process and monitor its progress. The marketing representative ensures that the product meets customer specifications throughout the design process, and the production and supplier representatives ensure that it can be produced at a reasonable cost. The entire process is much faster, consumes fewer resources, and has a higher chance of success.

In highly competitive environments where product functionality plays a critical role in the way firms compete, it is essential to launch new products on time. In such environments, the short product life cycles mean that firms do not have the luxury of designing products, discovering that they cost too much, and then redesigning them. They must design the product right the first time. The intense discipline target costing brings to the product development process is critical because it helps increase the probability that the cost of new products will be acceptable when they are launched.

When coupled to lean product development, target costing and value engineering maximize the probability that new products will both satisfy the customer and be profitable. There is some concern that target costing and value engineering slow down the product development process by requiring additional activities. To minimize delays, however, most firms perform the target costing process in parallel to other product development activities.

One outcome of the adoption of single-piece flow philosophies in product design is a considerable reduction in the time it takes to bring products to market. Product development cycles at Olympus, for example, have dropped from 10 years to 18 months. The outcome of this contraction is that it is harder for firms to develop sustainable, product-related competitive advantages, and this fact shapes the way lean enterprises compete. Competition between lean enterprises is based on the assumption that sustainable, product-related competitive advantages are extremely difficult to develop. These firms do not expect to be able to create mechanisms to avoid competition. Instead, they are forced to confront it and compete head on.

Confrontation is necessary because the fast reaction times of lean enterprises, the sophistication of the customer, and the rapid diffusion of technology inherent in the supply chains operate together to render product-related competitive advantages too fleeting to be considered sustainable. There is not enough time to educate customers to the positive attributes of the new product before other firms have me-too versions. This difference in the attitude of lean producers toward competition makes many of the lessons learned about competition between mass producers obsolete. Firms competing in industries dominated by lean enterprises must adopt the generic strategy of confrontation to survive. Unfortunately, confrontation is inherently expensive and therefore typically less profitable than strategies that manage to reduce or avoid competition.

THE THREE GENERIC STRATEGIES OF COMPETITION

Most of the existing literature on competition is based on the assumption that firms can develop sustainable, product-related competitive advantages and avoid competition by adopting the generic strategies of cost leadership and product differentiation. For example, the cost leader is able to offer products that are low in price and functionality by developing a sustainable cost advantage. This ability allows the cost leader to avoid competition by saying, "Don't compete with me. If you do, I'll drop prices even lower and render you unprofitable."

Similarly, differentiators offer products that have higher functionality than the cost leader's but that sell at higher prices. They develop unique products or services that closely satisfy customers' requirements, thereby isolating a section of the main market. This ability allows the product differentiators to avoid competition by saying, "This is my territory. I'm so good at what I do that attempting to compete with me is pointless."

The importance of competitive avoidance in Western strategic thinking is highlighted by its codification into the concept of strategic portfolio planning.[4] In this concept a firm tries to identify its "stars," that is, divisions that have successfully differentiated themselves or have become cost leaders. These divisions earn above-average returns that reflect their success at achieving a sustainable competitive advantage. To nurture new stars the firm protects fledgling units, the question marks, as they try to develop their sustainable competitive advantage. The ones that fail to create such an advantage, the "dogs," are either sold or liquidated. Those that have created a competitive advantage no longer considered sustainable are treated as "cash cows."

The contrary assumption—that competition is unavoidable—leads to the emergence of the generic strategy of confrontation. Firms that adopt this strategy do not attempt to become either cost leaders or differentiators. Instead, they try to keep their products ahead

---

[4] B. Heldey, "Strategy and the Business Portfolio," *Long Range Planning*, February 1977, p. 12.

of those of their competitors. It is important to understand that while these firms still try to differentiate their products, they do not expect to achieve sustainable competitive advantages; rather, they expect to achieve transitory ones. It is through transitory advantages that lean enterprises compete.

The Survival Triplet and Survival Zone

Three product-related characteristics, known as the survival triplet, play a critical role in the success of firms that have adopted the confrontation strategy. The survival triplet has an internal form that reflects the perspective of the producer and an external form that reflects the perspective of the customer (see Figure 1-7). Internally, the three characteristics are the product's cost, quality, and functionality. Externally, the characteristics are selling price, perceived quality, and perceived functionality. In the rest of this book the term "cost" will be used if the phenomenon being discussed is internal to the firm. If the phenomenon is external to the firm, the term "price" will be used for the cost/price characteristic. In addition, quality and functionality will be used to represent both internal and external views of those characteristics.

While the selling prices of products can be disconnected from costs temporarily, if the firm is to remain profitable in the long run costs must be brought into line with selling prices. Therefore, the survival triplet can be represented as cost/price, quality, and

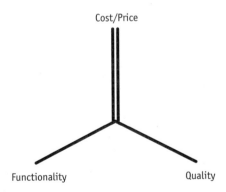

FIGURE 1-7. THE SURVIVAL TRIPLET

functionality. Here, cost/price is used to acknowledge that a long-term relationship exists between cost and price.

In the survival triplet price is defined as the amount at which the product is sold in the marketplace in arm's-length transactions. In the highly competitive markets in which most Japanese firms compete, the market sets selling prices, and cost is the value of the resources consumed to get the product into the hands of the customer. Cost includes all investment costs (such as research and development), all production costs, and all marketing and selling costs. Unlike price, it is not set externally but, like quality and functionality, has to be managed.

Quality is defined as performance to specifications. This definition of quality is narrower than the definition often used in literature on quality, in which quality is defined to include the ability to design a product that meets customer requirements (quality of design). This narrower definition allows quality and functionality to be treated as two separate characteristics.

Functionality is defined by the specifications of the product. It is not a single dimension but rather is multidimensional. When managers model competition using the survival triplet, they may find it beneficial to decompose functionality into a number of characteristics. For example, the firm may want to differentiate between the fundamental functionality of the product, such as the ability of a bulldozer to move earth, and the service functionality, such as the ability to guarantee 48-hour delivery of spare parts anywhere in the world. These two dimensions might be called product focus and customer focus. Such a differentiation will permit a richer modeling of the competitive environment and allow management to better understand the nature of the competition they face. Similarly, differentiating between the status that a product conveys (psychological functionality) and its pure performance (physical functionality) can be useful in explaining the success of products such as Rolex watches.

Each product a firm sells has distinct values for each of these three dimensions. Only products with values acceptable to the customer stand a chance of being successful. It is useful to define a survival zone for each product, identified by the gaps between the feasible and allowable values of the three dimensions of the survival triplet (see Figure 1-8).

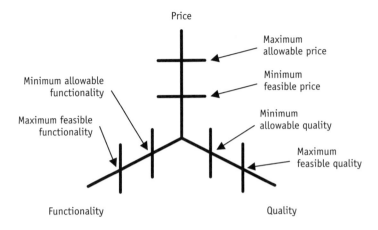

FIGURE 1-8. THE SURVIVAL TRIPLET FOR A PRODUCT

For quality and functionality, the minimum allowable level is the lowest value of each characteristic that the customer is willing to accept regardless of the values of the other two characteristics. Below a certain level of functionality, for example, few customers are willing to buy a product no matter how low the price or how high the quality.

The capabilities of the firm determine the maximum feasible values for quality and functionality. The maximum values are the highest values the firm can achieve without inducing significant penalties in the other characteristics. For example, above a certain functionality level, products will have quality problems and will need to be priced very high to make adequate profits. Low quality and high prices will result in too few customers being willing to buy the products. The maximum feasible value then represents the highest value the characteristic can have with respect to the other two characteristics and still have customers purchase the product.

The price characteristic is slightly different from the other two in that the customer determines the maximum allowable price and the firm determines the minimum feasible price. The maximum allowable price is the highest price the customer is willing to pay regardless of the values of the other two characteristics. The minimum feasible price is the lowest price the firm is willing to accept for the product if it is at its minimum allowable

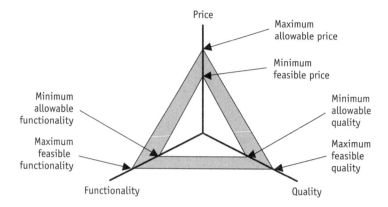

FIGURE 1-9. THE SURVIVAL ZONE FOR A PRODUCT

quality and functionality levels. While the customer views the critical characteristic as price, the firm views it as cost. The minimum acceptable profit at any price level transforms cost to price. The survival zone of a product can be identified by connecting maximum and minimum values (see Figure 1-9).

The Survival Zones for Mass and Lean Producers

In a market where only mass producers are competing, one mass producer occupies the cost leader position and the others occupy differentiator positions (see Figure 1-10). Theoretically there is no competition between the cost leader and differentiator as long as the quality and functionality gaps are sufficiently large to support the price gap. Thus, the cost leader launches products that are as close to the origin as possible but still inside their survival zones.

In contrast, the differentiator sells products that deliver higher quality and/or functionality than those of the cost leader, at a premium price. The differentiator launches products that are as far away from the origin as possible but still inside their survival zones. Since in practice functionality is not a single dimension, multiple differentiation strategies are possible, each designed to satisfy a different group of customers. It is this multidimensionality that allows multiple differentiators to coexist.

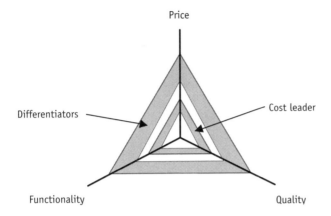

FIGURE 1-10. THE SURVIVAL ZONES OF THE COST LEADER AND
DIFFERENTIATORS

The emergence of the lean enterprise changes the shape of the survival zone. It increases the frontiers of functionality and quality considerably, but it simultaneously reduces their ranges (see Figure 1-11). The functionalaity improvements come about because the lean enterprise learns faster than its mass producer equivalent so it can increase the functionality of its products faster. However,

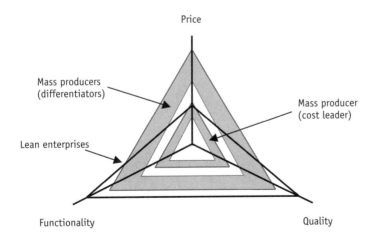

FIGURE 1-11. THE COLLAPSE OF THE COST LEADER AND
DIFFERENTIATOR SURVIVAL ZONES

the other lean enterprises with which it is competing catch up quickly, hence the reduced range. The same holds true for the other two dimensions of the survival triplet. In the narrow survival zones of the lean enterprise, becoming a differentiator is no longer possible. The bulk of the customers in the industry are unwilling to make significant trade-offs among the three dimensions, so there is simply not enough leeway for a firm to differentiate its products and sell them at a sufficient price premium to justify the increased costs.

All the firms offer products with similar high quality and functionality at a low cost. Thus, confrontation strategies apply when, in effect, all firms occupy the same competitive position. They are trying to attract the same customers with equivalent products. Because the firms are evenly matched on all three dimensions they have no choice but to compete head on and adopt a confrontational strategy. The outcome is a narrow survival zone (see Figure 1-12).

MANAGING THE SURVIVAL TRIPLET

Firms that adopt a confrontation strategy must become experts at developing low-cost, high-quality products that have the functionality customers demand. The cost, quality, and functionality expertise required for this result must be used to form a coherent strategy based on developing products with the right level of functionality

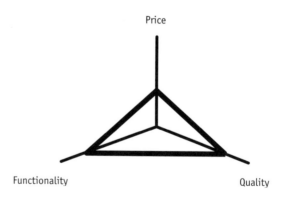

FIGURE 1-12. A CONFRONTATION SURVIVAL ZONE

and quality at the right price. Firms adopting the confrontation strategy must develop integrated quality, functionality, and cost management systems. It is the integration of these systems that allows many Japanese firms to respond so quickly to changes in economic conditions and to match the innovative products of their competitors.

Managing Functionality

A firm can compete in several ways using product functionality. It can, for example, accelerate the rate at which it introduces increased functionality. Olympus reduced the time it took to design a camera from 10 years for the OM10 to 18 months for a compact camera. Second, the firm can change the way it differentiates its products. For example, a firm that has historically differentiated its products based on higher functionality and quality for a higher price might suddenly launch products that offer different features for the same price. Olympus used this approach to increase market share when it introduced numerous products into the same price point using characteristics such as size and technical functionality. While horizontally differentiated products cost the same, they attract different customers. Finally, a firm can change the nature of the functionality of a product. For example, Swatch changed the functionality of its watches by making them a trendy fashion statement instead of just a way (attractive or otherwise) to tell the time.

Managing Quality

Quality is managed via the firm's total quality management (TQM) program. These programs have been well documented.[5] In many Japanese firms, the TQM programs have been so successful that they have increased the maximum achievable levels for the quality characteristic to such an extent that any additional improvements

---

[5] See, for example, P. B. Crosby, *Quality is Free*, New York: McGraw-Hill, 1979; P. Hauser and D. Clausing, "The House of Quality," *Harvard Business Review*, Vol. 66, No. 3, May–June 1988, pp. 63-73; and K. Ishikawa, *What is Quality Control? The Japanese Way*, Englewood Cliffs, NJ: Prentice-Hall, 1985.

are unlikely to be considered of value to the customer. When defects are measured in parts per million, individual customers are unlikely to encounter defects, let alone detect improvements in the defect rate! At the same time, the Japanese consumer demands such a high level of quality that even minimum acceptable levels are high by traditional standards. Consequently, the survival range for the quality characteristic is extremely small for most products, and quality has become a hygiene factor that can be ignored as long as it is under control. This does not mean that these firms have abandoned their TQM programs. Quality enhancements still result in internal benefits such as the ability to reduce the number of workers on the line, introduce new technologies faster, and reduce costs.

Managing Cost

Firms competing under confrontation must also manage costs aggressively. Lean enterprises reduce costs in three primary ways. The first is to manage the cost of future products, the second is to manage the cost of existing products, and the third is to harness the entrepreneurial spirit of the work force. While a number of factors determine the amount of energy a firm expends on each of these methods of cost reduction, three appear dominant. These factors are the competitive environment (in particular how the firm competes using the survival triplet), the maturity of the product's technology, and the length of the product's life cycle.

Managing All Three Dimensions Simultaneously

Typically, it is not necessary or advisable to expend equal effort on all three dimensions of the survival triplet. One dimension typically dominates the other two. The key to success is to select the appropriate rate of improvement for each of the three dimensions. For example, in a market where the customer is demanding increased functionality and is willing to pay for it, the most important dimension is functionality. In such a market, the firm that can increase the functionality of its products fastest (subject to price and quality constraints) will develop a competitive advantage. If, in contrast, the market is price driven, then the critical skill is cost reduction.

Failure to understand how to compete using the survival trip-let brings the risk of expending too much effort on one element of the triplet and not enough on the others. At the limit, the firm will fall outside the survival zone because it is too good at one or more elements of the triplet.[6] The critical point is that being too good is often as bad as being not good enough. This is not to say that firms do not have to improve on all three dimensions at once, but rather that different rates of improvement are required. When a firm's products fall outside the survival zone, it is imperative to get them back in as soon as possible.

The most important dimension frequently changes over time. For example, when the "bubble" burst and the yen strengthened against the dollar, many Japanese firms changed their most impor-tant dimension from functionality to cost.[7] These firms have devel-oped integrated systems flexible enough not to have to change when the most important dimension changes.

Surviving the Transition to Confrontational Competition

It is not clear that the number of competitors that can survive in a market dominated by mass producers is the same as the number that can survive in a market dominated by lean producers, nor is it clear that all firms can survive the transition. Each firm should evaluate its strengths and weaknesses and then decide whether its probability of success is high enough to make attempting the transition worthwhile. For weaker firms, it may be better to withdraw from the market than to die a slow, extended death trying to compete with insufficient resources. Firms that believe they are strong enough to survive need to develop the systems, organizational context, and culture required for the aggressive manage-ment of the survival triplet. All three are required and must be in harmony for the successful adoption of confrontation strategy.

---

[6] See "The Five Deadly Sins of Japan's Expanding 'High Tech Syndrome'," *Tokyo Business Today*, May 1992, pp. 34-36, for a discussion on "baroque engineering" and excessive functionality.

[7] Robert Neff et al., "How Badly Will Yen Shock Hurt?", *Business Week*, August 30, 1993, pp. 52-53.

To succeed, these firms must create an organizational context and culture that will support the integrated systems required to manage the survival triplet aggressively. Total quality management, product development, and cost management systems must be given a common theme.

The key point is that firms must learn to view the process of managing the survival triplet as a total systems solution, not a collection of independent techniques. For many firms undergoing the transition to confrontation, TQM and product development (time to market) systems function well, but the cost management systems fail to focus on the product design stage and are not adequately integrated into the firm's customer analysis program. Thus, of the three systems it is the cost management system that requires the most attention.

Once a firm is ready to adopt a confrontation strategy, it must carefully analyze competitive conditions to determine when its avoidance strategies are no longer appropriate. For many firms, delaying the adoption of confrontation strategy is appropriate because they can use the transition period to prepare themselves for the harsh realities of a confrontational environment. This decision involves a thorough analysis of the competitive environment, product life cycle, and technological maturity to determine the cost management techniques that are most likely to be beneficial. Each technique is applicable during a different stage of the product life cycle, and, depending on several factors, some techniques may be more or less beneficial for a given firm.

The nature of competition does not change immediately as firms in an industry begin to adopt lean enterprise practices. Instead, it changes gradually from competition avoidance to confrontation. This gradual change makes sense. Just as the cost leader generates extra profits by allowing the differentiators to create a price umbrella, the lean leader generates extra profits, in the early days, by allowing the mass producers to determine the positions of survival zones. As more companies become lean, however, the nature of competition changes, and the lean players begin to define the position of the survival zones. In time, confrontation becomes the dominant strategy and the remaining mass producers see their profits disappear as the ability to create and maintain sustainable competitive advantages based on the survival triplet becomes extremely rare.

Firms that have successfully developed sustainable competitive advantages in the old competitive environment may be tempted to make two mistakes. They may expend too many resources, first, trying to protect their existing (but no longer sustainable) competitive advantages and, second, trying to develop new ones. Unfortunately, all too often these firms will discover that all the resources committed to the extension or creation of sustainable competitive advantages were wasted: management has risked draining the firm of the very reserves it requires to adapt to confrontation.

The first step to surviving the transition is to accept that eventually the firm will have to adopt a confrontation strategy. This step is very difficult for many firms to accept because adoption of such strategies implies that profits eventually will fall below their historical levels. Firms that are used to being "profit maximizers" and are now adopting confrontation need to determine the time frame over which they are trying to maximize profits. If the firm uses too short a horizon, it risks adopting strategies that in the end will lead either to bankruptcy or, at least, to some very difficult years ahead. It may, for example, chase profits associated with niches that are no longer growing and therefore are attracting lower levels of competition than the main market. While the lower level of competition allows profits to be higher, the niche may not be attractive for the future.

If a mass producer stays in the main market, the only way it can maintain profit is either by increasing prices or by expending fewer resources on quality and functionality. For a while, this strategy can "succeed," but in the long run the underlying fallacies of such a strategy cannot be avoided. First, the firm is earning its high profit levels at the expense of customer loyalty. It is taking advantage of the loyal customers' willingness to accept products that are at the edge (the negative one) of the survival zone or even slightly outside it. Once this loyalty evaporates, the firm discovers that it has to catch up to survive. The other risk is that the extra resources the firm's competitors are expending on quality and functionality might give those firms the ability to shift the minimum acceptable values of quality and functionality at a speed that the profit maximizer cannot match. The profit maximizer's products rapidly fall out of their survival zones, and the firm is forced to catch up or exit the market.

SURVIVING IN
A CONFRONTATION MODE

Even when a firm makes a successful transition to confrontation strategy, it still faces the daunting task of surviving. Survival is achieved by creating an endless stream of temporary competitive advantages through aggressive management of the survival triplet. Every product must be inside its survival zone. Three factors drive changes in the position of a survival zone: changing customer preferences, the way in which competitors manage the survival triplet, and the firm's distinctive competencies. Changing customer preferences are important because they help determine which characteristics of the survival zone dominate. If the customer is demanding increased functionality, for example, and is not overly concerned with price, low-functionality and low-cost products are in trouble.

While customer preferences may change survival zones, a firm can influence customer preferences and thus survival zones by launching products with different trade-offs among the characteristics of the survival triplet. If a firm chooses to expend more resources on functionality than other firms in the industry, for example, it may gradually become the high-functionality (and presumably high-price) player. If the functionality of its products can be increased enough to make customers change their preferences, then the firm can change the position of the survival zones for the products it sells, forcing other firms in the industry to catch up.

The challenge for firms in a confrontational market is to determine when they can influence the customer preferences that drive their products. If a firm can identify one or more distinctive competencies, it should try to differentiate its products, even though the differentiation will be only temporary. Olympus, for example, launched the Stylus line of compact cameras because it believed it had a temporary advantage in its ability to develop very small cameras. While its competitors can now match the Stylus, Olympus still dominates the super-small compact camera market, demonstrating that, while first-mover advantages are reduced in confrontational environments, they are still real.

Becoming a Lean Leader

Firms become the lean leader by being better at managing the survival triplet than their competitors. There are two ways to become a lean leader. The first is to find a way to quantum jump the competition. When a firm achieves a quantum jump, it dramatically reshapes the survival zones of products. Topcon's development of near-infrared technology is an example of such a jump. Competitors are then forced either to catch up or go out of business. Unfortunately, it is extremely difficult for a firm to quantum jump its competitors because doing so requires identifying a new way to deliver enhanced product functionality. Like sustainable competitive advantages, a firm should always take advantage of such opportunities but accept that it is risky to rely solely on them to maintain profitability.

The second way in which a firm can become a lean leader is by achieving superior management of the survival triplet through the continuous improvement of its products. Achieving a leadership position in this way is difficult because all the firm's competitors are trying to improve at the same time. Usually each firm develops the leadership position in some of its products but not enough of them for customers to identify that firm as the leader.

A firm that is a lean leader and can maintain this position comes as close to developing a sustainable advantage as is possible in a confrontational market. The firm earns above-average profits or gains market share by sustaining the ability to generate a continuous series of temporary, not sustainable, advantages. This difference in the nature of the competitive advantage is real. Sustainable competitive advantages suggest static equilibrium, whereas temporary advantages suggest a dynamic status. Firms that try to avoid competition are forced to protect the status quo. They rarely take actions that will lead to head-on competition.

In contrast, when advantages are temporary, there is no status quo. Firms must seek continuously to find ways to develop advantages over their competitors so that they can successfully confront them. A market dominated by confrontation contains firms that are actively seeking to destroy the advantages of their competitors while creating new ones for themselves. Because the advantages are temporary, these firms frequently destroy their own current advantages to create new ones. After all, there is no point in trying to defend an advantage if it will disappear in the near future.

Lean Leader Strategies

A lean leader can use two strategies to take advantage of its position and superior profits. It can use its leadership position to accelerate the rate at which survival zones shift. In doing so, the firm attempts to leave its competitors behind so that their products fall outside their survival zones. This strategy increases the leader's market share and hence profitability. It forces every other firm into a catch-up mode. In the second strategy, the lean leader can allow the other firms in the industry to dictate the rate at which the acceptable values of each characteristic of the survival triplet shift. This strategy allows the firm to make superior profits because it can deliver products at lower cost through better management of the survival triplet than its competitors. Under this strategy profits increase, but market share remains the same.

While many factors determine which of the two strategies the lean leader adopts, the primary one is the relative technological capability of the firm in comparison to its competitors. If a leader has the most advanced products (i.e., has first-mover advantages) then it should adopt the first strategy. However, there is a risk in selecting this strategy. Continuously launching products at the limit of the firm's ability allows competitors to learn from the products and thus may accelerate the rate at which competitors can catch up. Firms that follow this strategy adopt a role equivalent to the differentiator's without achieving that status in the eyes of their customers. It is the first-mover advantages that generate the superior profits.

If the lean leader's first-mover advantages are small, it should think about adopting the second strategy. The extra profits generated by the firm's superior ability to manage the survival triplet reflect the firm's superior efficiency relative to its competitors. Firms using this strategy adopt a role equivalent to the cost leader's that allows differentiators to create a pricing umbrella from which to generate superior profits. However, as in the case of the first strategy, the firm does not achieve cost leadership status in the eyes of its customers.

If the lean leader's objective is to maximize profits (the traditional American objective), then the relative profits generated by the two strategies will determine which one is chosen. A profit-maximizing objective usually favors adoption of the second strategy. If the objective is to maximize market share (the traditional Japanese objective), then relative sales will determine which strategy is chosen. A growth objective usually favors the first strategy by launching products with higher quality and functionality than its competitors' products but not charging more for them.

Firms that are not leaders are jockeying to find ways to become leaders and could easily become leaders in the future. This ability reflects the dynamic nature of confrontational competition. No firms are "stuck in the middle" in Porter's sense. However, firms that do not aspire to leadership can carry out confrontation strategy in a third way. They can follow the leader. In doing this, the firm does little of the fundamental research and development required to introduce new products but instead uses value engineering and other techniques to quickly match its competitors' new products.

This follow-the-leader strategy works well if there are essentially two types of customers—trendsetters and copycats. Trendsetters buy new products as soon as they are available; copycats buy the products the trendsetters have already bought. If the delay between the trendsetter buying and the copycat buying is less than the time it takes the follow-the-leader firm to introduce its products, then first-mover advantages are small and the follow-the-leader strategy can be successful. If first-mover advantages are strong, which happens when trendsetters are more numerous than copycats, the follow-the-leader strategy will not succeed.

If the firms undertaking research and development into new products can find ways to develop expertise in the application of technology that cannot be learned through value engineering techniques, the follow-the-leader firms will fall behind. Alternatively, the firms undertaking research and development can license their new technologies to the other firms, and thus they can earn above-average returns.

SUMMARY

As more Western firms convert themselves to lean enterprises, the way they have to compete will change in critical aspects.

- Sustainable competitive advantages will become harder to achieve.
- The generic strategies of cost leadership and differentiation will begin to lose their effectiveness.
- Profit margins will shrink as competition becomes more confrontational.

Western managers must be sensitive to these changes because for many of the organizations the changes will be unavoidable. Managers who do not adjust their mode of competition accordingly will risk their firms. For example, managers who cling to the concepts of sustainable competitive advantage and who invest resources only in lines of business that they believe have such an advantage will discover that the number of investment opportunities they can identify is decreasing over time. Similarly, firms that cling to the traditional generic strategies of cost leadership and differentiation will discover that their ability to maintain those strategies is gradually eroding over time. They increasingly will be forced to retreat from their markets as lean competitors outmaneuver them.

A similar fate awaits firms that try to cling to their historical profit margins. The competition between lean enterprises is fiercer than between mass producers, and the overall profit margins are smaller. Retreating from products that have lower than usual profit margins will be successful only if other, higher-margin products can be identified. Often such products will not be available. If they are, there is the risk that chasing these high-profit products will lead to niches that, while initially highly profitable, eventually turn into technological dead ends.

Consequently, as the number of lean firms in an industry increases, managers should accept that they will have to change the way they compete. They will have to accept:

- the concept of transitory competitive advantages as opposed to sustainable ones,

- strategies based on confrontation instead of cost leadership and differentiation,
- the fact that niche strategies may no longer be successful,
- lower profit margins.

Firms that adopt a confrontation strategy will have to learn to manage the survival triplet in an integrated manner. In particular, once the firm achieves near parity with its competitors and its products are within their survival zones, it must choose how much energy to invest in improving each of the dimensions. Unfortunately, many Western managers have failed to understand the role of the survival triplet. They have adopted a rallying call of being the firm with the highest quality, lowest cost, and first-to-market products.[8] Such rallying calls are empty and simply distract the firm from choosing an appropriate strategy. These firms are not using the survival triplet properly. No firm can reasonably expect to be number one in all three elements of the survival triplet. Any firm that actually achieved this distinction would rapidly dominate its competitors. Furthermore, if it could sustain this advantage, it would become a monopoly because all its competitors would be bankrupt.

Many Western firms have adopted this best-in-all-three approach because they encountered Japanese competitors superior to them on all three counts, and to survive they had to improve on all three elements of the survival triplet simultaneously. The resulting struggle for survival caused many Western managers to lose sight of the critical fact that in most markets one element of the triplet is considerably more important than the others. What the managers failed to realize was that, once they were inside the survival zone, they no longer had to improve their performance along all three elements at the same rate. Instead, they had to learn to compete intelligently and select the rate at which they improved their performance on each element of the survival triplet.

To be successful, the cost management, quality, and time-to-market systems will have to be blended so that the survival triplet is managed as efficiently as possible. Managers will have to be sensitive

---

[8] Michael Hammer and James Champy, *Reengineering the Corporation: A Manifesto for Business Revolution*, New York: Harper Business, 1993.

to the changing importance of the three dimensions. The failure to be aware of this shifting can easily lead a firm to launch a new generation of products that fall outside their survival zones.

Another challenge will center around managing innovation and technology diffusion. First, the shift to multifunctional design teams will improve the firm's ability to integrate the three dimensions of the survival triplet but also risks a "sameness" in design. Managers will have to be creative in finding ways to take advantage of multifunctional groups without paying a price in terms of innovation. Management will also have to learn to control the way technology diffuses through supply chains. The technology sharing that occurs so freely in Japanese supply chains has the advantage of allowing firms to make innovations more rapidly, but at the same time it tends to lead to technological equivalence and hence confrontation.

The confrontation strategy is here to stay. It is a direct outcome of the emergence of the lean enterprise. It means more intense competition and lower profits. Adopting a confrontation strategy is not an excuse for senseless head-on competition. Instead, it is a deliberate strategy that acknowledges that lean enterprises do not compete in the same way as mass producers. Western managers must accept that in industries in which lean enterprises are becoming dominant, only those firms that adopt confrontation strategy and aggressively manage the survival triplet will survive.

CHAPTER 2

THE ROLE OF COST MANAGEMENT IN CONFRONTATION STRATEGY

INTRODUCTION

Firms that adopt a confrontation strategy must learn to manage costs as aggressively as possible. It is only if firms understand the importance of cost management that they can manage the survival triplet correctly and succeed in using a confrontation strategy.

Cost management plays an important role in the success of many Japanese firms. Japanese firms that have extended experience with the confrontation strategy form of competition have developed sophisticated cost management systems to help them manage costs aggressively. Eight cost management techniques are used to reduce costs aggressively. The first three of these techniques focus on managing the cost of future products. The next three manage the costs of existing products. The final two harness the entrepreneurial spirit of the workers.

MANAGING THE COST
OF FUTURE PRODUCTS

Managing the cost of future products is important because it is the only way to ensure that future products will be profitable when launched. Firms have developed three feedforward cost management techniques to help them manage the costs of future products: *target costing, value engineering,* and *interorganizational cost management* systems (see Figure 2-1). These systems are particularly important because evidence shows that most of a product's costs are designed in. Once the product enters production it is too late to make significant cost reductions without essentially redesigning it. The three feedforward techniques are designed to ensure that, if possible, costs are not designed into products in the first place.

Target Costing

Target costing is the discipline that ensures that new products are profitable when they are launched. It has three major steps. The first is to determine a product's target selling price and target profit margin so its allowable cost can be established. The second is to set an achievable product-level target cost. The third is to decompose the product-level target cost down to the component level so the purchase price of the components can be determined.

A product's target cost is arrived at by subtracting its target profit margin from its target selling price and adjusting for the cost-

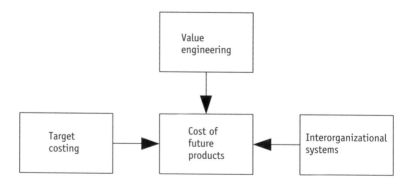

FIGURE 2-1. MANAGING THE COST OF FUTURE PRODUCTS

reduction capabilities of the firm and its suppliers. The target selling price of a new product is decided primarily from market analysis. The target profit margin is based on corporate profit expectations, historical results, and competitive analysis.

The critical factor that distinguishes target costing (as opposed to other approaches to managing the costs of new products) is the intensity with which the cardinal rule is applied: the target cost can never be exceeded. Without the application of such a rule, target costing systems typically lose their effectiveness. In practice, the cardinal rule may be broken at times, but the conditions must justify it and specified procedures must be followed to authorize it. Design engineers are definitely not allowed to make decisions of the type, "If we just add this feature, the product will be so much better (and only cost a little more)."

Once the product's target cost has been established, the next step is to decompose that cost to the component level. The chief engineer usually is responsible for allocating the target cost among the major functions of the product. He defines the main themes of the new model. For example, for an automobile it might be a quieter but sportier ride. Once the target costs of the major functions are set, the next step is to identify the target cost of the components they contain. These component-level target costs become the purchase prices of the externally acquired components. That is, they are the suppliers' selling prices as set by their customers.

Value Engineering

Value engineering (VE) is a systematic, interdisciplinary examination of factors affecting the cost of a product so as to devise a means of achieving the specified purpose at the required standard of quality and reliability at the target cost. VE, like target costing, is applied during product development. VE is a multidisciplinary, team-based approach. Teams typically are drawn from multiple functional areas, including design engineering, applications engineering, manufacturing, purchasing, and sometimes even the firm's suppliers and subcontractors.

VE is an organized effort to analyze the functions of goods and services so a firm can find ways to achieve those necessary functions and essential characteristics while meeting its target costs. VE plays a critical role in the cost management of future products because it helps the firm manage the trade-off between functionality and cost. As this trade-off is critical in many competitive environments, it is not surprising that many Japanese firms have strongly embraced VE practices.

VE requires identification of each product's basic and secondary functions and analysis of the functions' values. A basic function is the principal reason for the existence of the product. For example, an automobile's basic function is to provide transportation. The secondary functions are outcomes of the way the designers chose to achieve the basic function, for example, the heat and pollution generated by the auto engine.

An important aspect of Japanese VE programs is that their objective is not to minimize the cost of products, but to achieve a specified level of cost reduction (the product's target cost). This difference between the Japanese and Western approaches is significant because designing to a specified low cost appears to create more intense pressure to reduce costs than designing to an unspecified minimum cost.

Interorganizational Cost Management Systems

The intense pressure to become more efficient causes many firms to try to increase the efficiency of their suppliers of raw materials and components by developing interorganizational cost management systems. These systems emerge because it is no longer sufficient to be the most efficient firm; it is also necessary to be part of the most efficient supply chain.

To achieve this objective, many Japanese firms are opting to blur their organizational boundaries in numerous ways. Organizational blurring typically occurs when information critical to one firm is possessed by another firm further up or down the supply chain. The two or more firms then create relationships that share organizational resources, including information that helps improve the efficiency of the interfirm activities. Mechanisms for information sharing include joint research and development projects, placing employees of one firm in others.

There are two primary approaches to interorganizational cost management systems—chaining target costing systems and bringing together engineers from all the firms involved in the design and manufacture of a product. Target costing systems can be used to set the purchase price of externally acquired components. If the supplier has a target costing system, the customer's target costs for components becomes the target selling price for the supplier. Thus a chain of target costing systems can extend across suppliers. Each system communicates the competitive pressures to the next supplier in the chain. The limitation of such a chain is that it creates local optima.

Sometimes a more efficient solution can be found by bringing together design engineers from every firm involved in the design and manufacture of a component. Engineers from each firm typically will have expertise in different technologies. It is only when these engineers can pool their knowledge that innovative solutions sometimes can be identified.

MANAGING THE COST
OF EXISTING PRODUCTS

Managing the cost of existing products is the second way costs are reduced. It includes three feedback techniques: *kaizen costing, product costing* systems, and *operational control* (see Figure 2-2). Their primary purpose is to create cost-reduction pressures on the production processes. The need for both feedforward and feedback cost-management techniques reflects the different objectives of the

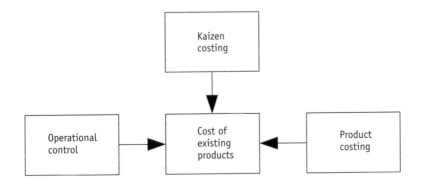

FIGURE 2-2. MANAGING THE COST OF EXISTING PRODUCTS

two types. Feedforward techniques focus on reducing costs through more efficient product design, and feedback techniques focus on reducing costs through more efficient production. Together, the six techniques are used to create a continuous downward pressure on costs across the entire life cycle of the firm's products.

Product Costing

Given the importance Japanese firms attach to cost management, their cost systems might be expected to be technically advanced and capable of reporting extremely accurate product costs. In particular, activity-based costing (ABC) systems might be expected to be either emerging or already in place. Despite this expectation, many of the systems we encountered were relatively traditional, and equivalent systems can easily be found in Western firms. Not all the systems we encountered, however, were conventional in their design. Several contained very innovative designs that reflected the economic realities of the competitive environments the firms faced, the nature of their production processes, and the types of decisions that management considered most important. For example, some systems reported product-line as opposed to product costs.

The dominance of product-line as opposed to product-level decisions reflected the fact that some of the firms studied sold carefully designed families of products. The underlying marketing rationale of such a strategy is to break the market into a number of distinct segments, each containing a large number of customers. Each of these segments demands a different primary function or set of functions for the product. In the case of compact cameras, it might be the focal length of the lens or its zoom capabilities. For a car it might be the type of engine, the comfort of the ride, and the appearance. Using these primary functions, the firm designs a set of products that will satisfy virtually every customer. Customers then choose the product that best satisfies their requirements. By designing a complete product line, the firm gives the consumer no reason to look at competitive offerings. In contrast, if the customer is not satisfied with the product offering, then he or she will go elsewhere. Only if that firm fails to keep the customer satisfied will the original firm have an opportunity to win the customer back.

Leaving a "hole" in the line by choosing not to fill one of the segments because the product is unprofitable is unacceptable because everyone trading up or wishing to buy the missing product will simply switch to another firm. Consequently, any estimate of the profitability of a product must include the future profits from subsequent sales to customers who buy "unprofitable" products. Most firms believe that all high-volume segments are profitable in the long term, so the firms provide products accordingly. Given this strategy, many firms did not consider product-mix management particularly important at the individual-product level. All significant product decisions were made at the product-line level, for example, whether to manufacture all black and white televisions off-shore or, as was the case with Topcon, whether to cease manufacturing all cameras. Only rarely did these firms make discontinuance decisions about individual products. As a reflection of the diminished importance of product-level decisions, some of the firms studied implemented cost systems that could report only product-line costs.

Operational Control

Operational control requires holding people responsible for the costs they control and determining how well they manage them. The two primary techniques of operational control are the establishment of responsibility centers and variance analysis. For an individual to be held responsible for a cost, that cost must be assigned directly to the center over which that person has control. If indirect cost assignments are used, then it is impossible to hold the person responsible for any apparent changes in the level of resource consumption; that person cannot be held responsible because there is no way to know if apparent changes in resource consumption are due to distortions in the indirect assignment process or to an actual change in the level of consumption. At the firms studied, Japanese managers were well aware of the problems of trying to hold individuals responsible for indirect costs. For example at Mitsubishi Kasei (an industrial chemical manufacturer), the existing cost system suffered from a very high level of allocations. These allocations made it almost impossible to assign responsibility for costs. This inability was considered critical and was the primary motivation to design a new system that would provide increased cost control.

The traditional use of variance analysis is to monitor how well the responsible manager is keeping control over costs. The aim is to ensure that the budget is achieved. Some of the Japanese firms studied used variance analysis in this way. For example, the variance analysis performed at Komatsu is taken straight from the textbooks with only minor variations. Typically, variance analysis is used to determine if task performance is adequate.

Other firms used variance analysis to monitor how well they were achieving their kaizen objectives. In environments with kaizen costing programs, the standards used to generate variances usually reflect the expected improvements due to continuous improvement activities.

At several of the firms, the standard cost system and the kaizen program were integrated so that the two supported each other. The simplest approach used to achieve this objective was to embed the anticipated kaizen improvements into the standards.

Kaizen Costing

Kaizen costing stands for continuous improvement. It is the application of kaizen techniques to reduce the costs of components and products by a prespecified amount. The difference between target and kaizen costing focuses on the point in the life cycle at which the techniques are applied and their primary cost-reduction objective. Target costing is applied during the design stage of the product life cycle. It achieves its cost-reduction objective primarily through improvements in product design. In contrast, kaizen costing is applied during the manufacturing stage of the product life cycle. It achieves its cost-reduction objectives chiefly through increased efficiency of the production process.

There are two types of kaizen costing, product-specific and general. Produce-specific kaizen costing is applied under two conditions: first, when a product is launched above its target cost and second, when an existing product's profitability is threatened by price reductions. In both cases, engineering teams are created to find ways to reduce costs without altering product functionality, for example, by replacing metal components with plastic ones or reducing parts count by shifting to more integrated components. The second type of kaizen costing does not focus on individual products but on making the firm's production processes more efficient. In environments in which products have short lives, production processes are often extended across several generations of products, which reduces the cost of the production process and can lead to long-term savings.

HARNESSING THE ENTREPRENEURIAL SPIRIT

Another way to create pressure to reduce costs is to implement cost management techniques that harness the entrepreneurial spirit of the work force. These techniques are fundamentally different from those described above because they do not focus on either the product or the production process but on motivating the work force. There are at least two cost management techniques for accomplishing this objective. The first technique creates *pseudo micro profit centers* from cost centers, and the second technique converts the firm into a large number of *real micro profit centers* (see Figure 2-3).

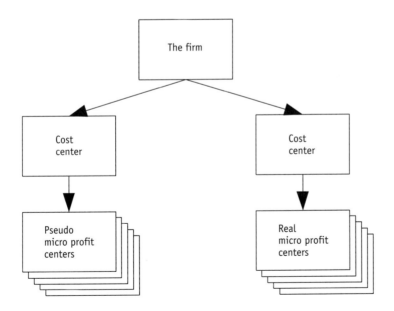

FIGURE 2-3. HARNESSING THE ENTREPRENEURIAL SPIRIT

Pseudo Micro Profit Centers

Firms that use the first technique convert cost centers into profit centers and change work group leaders from cost center managers to business managers. The development of this technique is motivated by the belief that the way in which people view their responsibilities sometimes can be as important as the responsibilities themselves. The technique is useful for four primary reasons:

- By using profit as the performance metric, work groups can get a better feel for the impact of their contributions on company performance.
- The creation of profit centers provides management with a mechanism to reward individuals publicly and thus reinforce behaviors that lead to increased profits.
- Because profits are a universal metric, each work group can evaluate its own performance and compare it to that of other groups.
- The creation of profit centers can revitalize the firm's cost management systems.

Real Micro Profit Centers

The second technique consists of breaking the firm into a collection of autonomous small enterprises that must be profitable to survive. Under the approach used by Kyocera, a large number of profit centers, called amoebae, are created. They are not independent firms but highly independent pseudo firms that are responsible for selling products both internally and externally. In contrast, the approach used by the Taiyo Group creates separate legal entities, each responsible for several products. At the heart of both approaches is the fundamental assumption that small firms are inherently more efficient and effective than large firms—they do not require an expensive and ineffective bureaucracy, and they can react quickly to changes in competitive conditions. Both firms believe the ability to reduce or control the growth of bureaucracy is a major mechanism to control costs.

SUMMARY

When firms adopt a confrontation strategy, aggressive management of the survival triplet becomes critical. For such firms, a highly effective cost management program is a necessity, not a luxury. Reflecting the importance of cost management to their strategy, many Japanese firms have developed integrated cost management programs that create a discipline throughout the firm to reduce costs across the entire life cycle of the firm's products.

These programs consist of six product- and production process-related techniques and two that harness the entrepreneurial spirit. Of the six product- and process-related techniques, three are feedforward techniques designed to help manage the costs of future products. They are target costing, value engineering, and interorganizational cost management systems. The next three techniques are feedback techniques designed to help manage the costs of existing products. They are kaizen costing, product costing, and operational control. The last two techniques, pseudo micro profit centers and real micro profit centers, focus on harnessing the entrepreneurial spirit of the work force.

Japanese cost management programs are designed to affect all aspects of the economics of manufacture. They influence the supply of purchased parts, the design of the products, and the manufacture of the products. At each stage of the production and delivery process, Japanese firms have developed techniques to reduce costs.

For a given firm, the effectiveness of the various cost management techniques appears to depend on several factors, including the competitive environment, the maturity of the technologies used in the products, and the length of the product life cycle. The role of these three factors in determining the effectiveness of the six cost management techniques is shaped primarily by how the firm is using the survival triplet.

Feedforward techniques are particularly effective when firms are competing primarily on the functionality of their products. Such firms must continuously introduce new products with enhanced functionality to survive. Typically they rely on the latest technologies and reduce product life cycles to a minimum. The feedback techniques are particularly effective when firms are competing primarily on price. Such firms usually are dealing with more mature technologies and longer life cycles and to reduce costs to a minimum have to make their production processes as efficient as possible.

Firms that are competing primarily on quality probably will rely on a mixture of the techniques depending on whether they are using technology to enhance quality or improve production processes. Firms relying on new technologies will benefit from target costing, while those relying on production process improvements will benefit from kaizen costing.

CHAPTER 3

THE RESEARCH PROJECT

INTRODUCTION

In the summer of 1994, the Institute of Management Accountants and the authors launched a joint research project to document and analyze the cost management practices at Japanese firms. This project was a continuation of one initiated by Robin Cooper that had culminated in the publication of 23 Harvard Business School cases on the cost management practices at 19 Japanese firms and the book *When Lean Enterprises Collide: Competing Under Competition.*

The primary objectives of the continuation project were to: (1) complete the study of the cost management techniques identified in the original study, (2) develop additional case studies on Japanese cost management practices, (3) present the cases in a more accessible form to managers, and (4) create an in-depth analysis of the practices at the firms so that the following questions could be answered:

- What is the technical nature of each cost management technique?
- How is it applied in practice?
- What are the implications for Western managers?

The Research Approach

The researchers visited 25 firms to collect information about the issues that led management to implement advanced cost management systems, the insights obtained from the information reported by those systems, and the actions taken based on those insights. After the research team had reviewed the findings from all the research sites, the team analyzed and synthesized the information. They developed a framework for understanding the practice of each cost management technique encountered and did an in-depth analysis of that practice.

The research process consisted of five major steps:

- forming the project committee and research team,
- selecting the research sites,
- visiting the firms and collecting information about the cost management techniques used and how they were applied,
- writing the cases, and
- analyzing and synthesizing the findings.

Forming the Project Committee and Research Team

At the outset of the project, the IMA established a project committee to oversee the entire project. Its members were Robert C. Miller, Boeing Corporation, and Hank Davis, Dodge Rockwell Automation. The research team, consisting of Robin Cooper and Regine Slagmulder (plus coauthors), selected the research sites, visited the sites, researched the individual cases, prepared and cleared the cases, analyzed the research findings, and prepared a synthesis of those findings. Robin Cooper provided overall management of the project, including the liaison with the IMA and the project committee.

Site Selection — The Series

All the research sites documented in this series are located in Japan. Japanese firms were chosen because they have extensive experience with competition among lean enterprises. This experience has given them time to develop mature cost management systems required to compete in a confrontational mode. The 31 cases document the cost management practices at 25 manufacturing organizations (listed in the Appendix). These firms are drawn from a number of different industries and vary in size from very large to quite small.

To ensure that the sample for the entire project included a broad range of cost management techniques in different industries, the research team established four site selection criteria.

- All of the sites had to have been lean enterprises for some time. This criterion reflected the objective of the research project, to study the cost management systems of lean enterprises.
- The firms had to have well-developed and frequently updated cost management systems. The objective of this criterion was to ensure that the cost management systems documented were not out of date but reflected best practices.
- The firms were to be chosen from a cross section of manufacturing industries (heavy manufacturing, light manufacturing, and process). This criterion was established to explore both the range of cost management practices in Japanese lean enterprises and the range of application of those techniques.
- When possible the firm had to have a reputation for being well managed and having innovative cost management systems. This objective helped ensure that the best Japanese cost management practices would be documented.

Approximately 50 firms were contacted in a variety of ways. Japanese academics and professional contacts of the researchers identified some of the firms. Others were chosen because articles and cases had been written about them. Managers at firms in the sample pointed out a third group of firms as having highly innovative systems. Finally, some were contacted directly based on their general reputation for being well managed and having innovative products and management systems.

The final selection was based on whether each company satisfied the site selection criteria, agreed to become a research site, and was willing to allow the field research teams to complete the site visits within a reasonable time frame. Because of the selection method, the company sample is neither random nor necessarily representative of the population of lean enterprises in Japan.

SITE SELECTION — THIS VOLUME

The sample firms selected for this particular volume were identified in several ways. Japanese academics indicated some as having excellent target costing and value engineering systems. Michiharu Sakurai of Senshu University identified Nissan and Komatsu, Takao Tanaka of Tokyo Keizai University named Toyota, and Takeo Yoshikawa of Yokohama National University identified Isuzu. A Japanese executive of Sony chose that firm. Finally, Olympus and Topcon were contacted directly based on their reputation for innovative and competitive products. In both cases, a letter was sent to the firm's president requesting involvement in the research project. Both companies agreed to participate and selected their target costing system as the most noteworthy part of their cost management systems.

The seven firms included in the target costing and value engineering sample are described below.

Isuzu Motors, Ltd.

Isuzu originated in 1916 with Tokyo Ishikawajima Shipbuilding and Engineering Company, Ltd. manufacturing automobiles. Based on units produced, Isuzu was the ninth-largest automobile company in Japan in 1992 with 10 large domestic competitors. The 4% market share of Isuzu did not reflect the firm's specialized market strength in trucks and buses due to the higher-volume passenger car market. Isuzu had 10% market share in heavy- and light-duty trucks and 11% share in the bus market.

Komatsu, Ltd.

Founded in 1917 as part of the Takeuchi Mining Company, this firm was one of the largest heavy industrial manufacturers in Japan. It was organized in three major lines of business—construction equipment, industrial machinery, and electronic-applied products—which accounted for 80% of total revenues. The remaining 20% consisted of construction, unit housing, chemicals and plastics, and software development. These products together generated revenues of ¥989 billion and net income of ¥31 billion in 1991, making Komatsu a large international firm. Since 1989, the company has been aggressively diversifying and expanding globally.

Nissan Motor Company, Ltd.

This firm was founded in 1933 and considered itself the most highly globalized of the Japanese automobile companies with 36 plants in 22 countries and marketing in 150 countries through 390 distributorships and over 10,000 dealerships. In 1990, Nissan was the world's fourth-largest automobile manufacturer, producing just over three million vehicles, filling approximately 10% of the world's demand for cars and trucks. Nissan had a stated policy of globalization through a five-step process that emphasizes localization of production, sourcing, research and development, management functions, and decisions.

Olympus Optical Company, Ltd.

As part of Olympus, Olympus Optical Company manufactured and sold opto-electronic equipment and other related products. Originally called Takachiho Seisakusho, Olympus was founded in 1919 as a producer of microscopes. Major product lines were cameras, video camcorders, microscopes, endoscopes, and clinical analyzers. By 1995, Olympus was the world's fourth-largest camera manufacturer with consolidated revenues of ¥252 billion and ¥3 billion in net income.

Sony Corporation

Sony, one of the world's largest electronics companies, started as Tokyo Telecommunications Research Institute and generated revenues in its earlier years by repairing broken radios and manufacturing short-wave converters. The company's first really successful product was Japan's first magnetic tape recorder in 1950. The company continued to grow rapidly, and by 1960 it became a truly international firm with Sony Corporation of America and Sony

Overseas, S.A., in Switzerland, followed by Sony UK and Sony GmbH in 1968 and 1970 respectively. The first Sony Walkman was introduced to the Japanese public in July 1979, and by 1994, Sony had sold more than 120 million units.

Topcon Corporation

Topcon was founded in 1932 as the Tokyo Optical Company, Ltd. It diversified along its core competencies in advanced optics and precision equipment processing. By 1992, Topcon sold four major product lines: surveying instruments, medical and ophthalmic instruments, information instruments, and industrial instruments. The surveying instruments business unit contributed approximately 36% of sales; medical and ophthalmic instruments, 28%; information instruments, 13%; and industrial instruments, 23%. Topcon specialized in high-technology, high-margin, and low-volume products. To be able to rely continuously on high technology for profit, Topcon invested heavily in research and development.

Toyota Motor Corporation

Toyota Motor Corporation started as a subsidiary of the Toyota Automatic Loom Works, Ltd. It was founded in 1937 as the Toyota Motor Co., Ltd. It changed its name to the Toyota Motor Corporation in 1982 when the parent company merged with Toyota Sales Co., Ltd. In 1993, Toyota Motor Corporation (Toyota) was Japan's largest automobile company. It controlled approximately 45% of the domestic market. Over the years, Toyota changed from a Japanese firm into a global one. In 1993, a considerable part of the firm's overseas markets were serviced by local subsidiaries that frequently designed and manufactured automobiles for local markets.

Visiting the Sites and Writing the Case Studies

Managers, design and manufacturing engineers, and blue-collar workers who were actively involved in applying target costing and value engineering techniques at the seven firms were interviewed.

In-depth interviews with three to five persons in each firm were held in English with translator support as appropriate. These persons were responsible for one or more target costing projects. Job titles included general manager of product planning, manager of corporate planning, chief engineer, and senior manager of group accounting. The total site visits lasted 20 days. The average visit to each firm was three days. Typically, the firm was visited initially for two days, and a draft of the case was prepared. Follow-up visits took place to clear up any major outstanding issues and to obtain agreement to sign the final draft.

Copious notes and tape recordings of the interviews were the basis for research cases of approximately 5,000 words each. These cases were sent to the contact manager in each firm for review. The first draft of the cases contained numerous questions that the authors could not answer from their tape recordings and notes.

The cases typically went through two to three revisions before being cleared. It took 12 to 18 months to clear each case. When necessary, the questions and appropriate textual portions of the case were translated into Japanese so that managers with inadequate English skills could answer the questions and review the text for accuracy. In a typical clearance procedure approximately 60 questions were answered and about one-third of the case was rewritten or amended in some way. While the majority of these changes related to the author-initiated questions, others were corrections to the drafts made by the reviewing managers. If necessary, competitive or sensitive information was disguised at the request of the participating company. After the reviewing managers at the firms had answered the questions and were satisfied that the cases were factually correct, the cases were used as a basis for writing this volume. Ultimately, each company signed a representation letter authorizing release of the final document.

ANALYZING AND SYNTHESIZING THE CASE STUDIES

Once the cases were completed, the outputs of target costing and the processes used to achieve them were analyzed. Three major areas in the target costing process: price-driven costing, product-level target costing, and component-level target costing were identified. The output of the price-driven costing process is an allowable cost for each future product. The allowable cost is the cost at which the product must be manufactured if it is to achieve its profit objective when sold at its target price. The steps required to generate allowable costs are described in Chapter 5. The allowable cost does not take the capabilities of the firm and its suppliers into account, and therefore it sometimes has to be increased to make it achievable.

Chapter 6 describes the process of generating achievable target costs, and Chapter 7 discusses value engineering, the primary technique for achieving target costs. The process of developing component-level target costs is described in Chapter 8. Chapter 9 deals with the factors that influence the target costing process. Chapter 10 uses a case study of a hypothetical company to show how target costing and value engineering are applied. Finally, Chapter 11 provides a checklist of factors that potential adopters of target costing and value engineering should assess.

PART 2

TARGET COSTING
AND VALUE ENGINEERING

CHAPTER 4

An Overview of Target Costing

and Value Engineering

Introduction

Target costing is primarily a technique to manage the future profits
of the firm. It achieves this objective by disciplining the product
development process—identifying the cost at which the product
must be manufactured if it is to achieve its target profit margin
when sold at its target selling price. Once this target cost has been
established, value engineering (VE) is used to find ways to improve
the product design so that the target cost can be achieved.

Target Costing

The process of target costing begins when a product is first conceptualized and ends when it is released for mass production. Target
costing plays a critical role in managing costs because once a product is designed, most of its costs are committed. For example, the
number of components, the types of materials used, and the time it

takes to assemble are all determined primarily by the design of the product. Some authorities estimate that as much as 90% to 95% of a product's costs are designed in; that is, they cannot be avoided without redesigning the product (see Figure 4-1). Consequently, effective cost management programs must begin at the start of the design phase of a product's life cycle.

Target costing is a *structured approach to determine the life-cycle cost at which a proposed product with specified functionality and quality must be produced to generate the desired level of profitability over its life cycle when sold at its anticipated selling price.*[1] Target costing makes cost an input to the design process, not an outcome of it. By estimating the selling price of a proposed product and subtracting the desired profit margin, the cost at which the product must be manufactured—its target cost—can be identified. The key is to design the product so that it satisfies customers and can be manufactured at its target cost.

Target costing is as much a tool of profit management as it is of cost management. In Japan, where the lean enterprise evolved, firms have learned to view target costing not as a stand-alone program but as an integral part of the product-development process. These firms have developed target costing into a powerful mechanism to bring the competitive challenge of the marketplace through the organization to the product designers to ensure that only profitable products are launched. They use it to create a discipline that coordinates the labor of disparate participants in the product-development effort, from designers and manufacturing engineers to market researchers and suppliers. Target costing helps harmonize the product-design effort by creating a common language of design.[2] This common language is critical if the marketing, design, and manufacturing functions and the firm's suppliers are to communicate effectively with one another. In firms that don't practice target costing, all too often the participants develop idiosyncratic languages based on the way they view the product (see Table 4-1).

---

[1] Target costs should include any costs that are driven by the number of units sold. For example, if the firm accepts responsibility for disposing of a product at the end of its useful life, these costs would be included in the target cost. See Robin Cooper and B. Chew, "Control Tomorrow's Costs Through Today's Designs," *Harvard Business Review*, January-February 1996, pp. 88-97.
[2] Ibid.

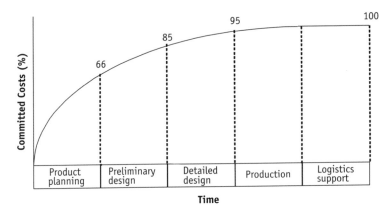

Adapted from B. S. Blanchard, DESIGN AND MANAGE TO LIFE-CYCLE COST, Portland, OR: M/A Press, 1978.

FIGURE 4-1. COMMITTED COSTS

For example, customer requirements are usually expressed in terms of product attributes, product design features in terms of functionality, manufacturing steps in terms of subassemblies, and supplier specifications in terms of components. Unfortunately, these disparate languages make it difficult for the functions to communicate effectively with each other.

While each of these languages might be perfectly adequate for use within a function, none of them addresses overall profit targets. Consequently, there is no single context within which everyone involved in the product-design process can work; for example, a

TABLE 4-1. DIFFERENT VIEWS OF THE PRODUCT

Customer View	Designer View	Manufacturing View	Supplier View
Performance	Engine cooling system	Radiator/fan/motor subassembly	Valves
Safety		Wheel rim/tire/brake subassembly	Pistons
Styling	Brakes	CD player/wiring harness/speaker subassembly	Crankshafts
Maintenance	Sound system		CD player
			Amplifiers
			Speakers

product designer assigned to a particular subsystem can't get a clear answer to the questions of how a subsystem is valued by customers or what a subsystem will cost. Target costing provides a context within which everyone involved in the product-design process can work effectively together by using a common language. Target costing creates an overarching product-design strategy that focuses the design team on the ultimate customer and the real market opportunities.

The Structure of the Target Costing Process

Target costing, to be effective, must be a highly disciplined process (see Figure 4-2). The discipline of target costing starts by forcing alignment with the marketplace and requiring a new level of specificity about what customers want and what price they are prepared to pay. Market analysis helps determine the location of the survival zones of new products. It plays a critical role in shaping the *market-driven costing* portion of target costing by determining allowable costs. Target costing uses these allowable costs

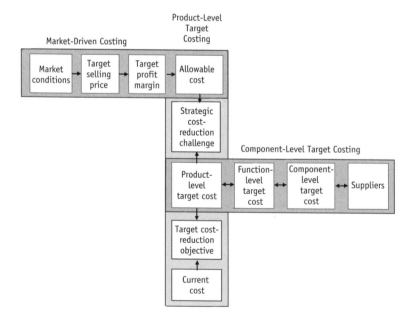

FIGURE 4-2. THE TARGET COSTING PROCESS

to transmit the competitive cost pressures the firm faces to the product designers.

The product designers must ensure that new products are inside their survival zones when launched. *Product-level target costing* disciplines and focuses the creativity of the product designers on achieving the cost aspect of this objective. Including product functionality and quality in the target costing process ensures that the designers consider all three characteristics of the survival triplet when designing products. Once product-level target costs are established, they are decomposed to the component level, thus transmitting the cost pressures the firm faces to its suppliers. Suppliers, in turn, must find ways to design and manufacture the firm's externally sourced components so that they can make adequate returns. Thus, *component-level target costing* helps discipline and focus the creativity of the suppliers in ways beneficial to the firm.

The Seven Key Questions of Target Costing

The essence of target costing can be captured in the following seven questions.

- What are the firm's long-term sales and profit objectives?
- Where will the new product's survival zone be when it is launched?
- What is the target profit margin?
- What level of cost reduction is realistic?
- How can we achieve this cost reduction objective?
- Are there extraneous circumstances that allow the target cost to be relaxed?
- How can we distribute the cost reduction among the components?

The first question deals with the firm's strategic objectives. Top management begins the target costing process by setting the long-term sales and profit objectives of the firm. To support these objectives they establish a long-term product plan by undertaking detailed analyses of customer and competitive trends. The aim of these analyses is to identify the mix of products considered most likely to enable the firm to achieve its long-term objectives.

The second question focuses the level of the analysis on the individual products in the long-term product plan. Companies that rely on target costing expend considerable resources on market analysis to determine how customer preferences and competitor products are evolving over time. The objective is to launch products that satisfy customers more than the competitors' equivalents, at the right price and at the right time, that is, are inside their survival zones. Very close interactions with customers often are required to enable the firm to predict the location of future survival zones.

For example, Olympus regularly assigns product engineers to spend time interacting with customers in camera stores. By gathering this type of firsthand market information, the company aims to fine-tune its ability to understand what the market expects. One critical output of this stage of the analysis is the target selling price of the new product.

The third question deals with the profit margin the new product must earn. The typical starting point for determining the target profit margin for a new product is the margin similar products have earned in the past. If all else remains the same, this is the margin that will be used to determine target costs.

The relative strength of competitive offerings can cause the historical margin to be modified. If the firm believes that its new products are superior to those of its competitors, then it might increase the actual selling price and hence profit margin. Alternatively, if the new products are functionally inferior, then prices and hence profit margins may have to decrease. Many firms use computer simulations to help them estimate overall firm profitability based on individual target profit margins.

The product's target cost is not always simply the target selling price minus the target profit margin. Sometimes adjustments have to be made for the capital intensity of products and the ability to reduce manufacturing costs during the product's life. When up-front development costs and the capital investment required to launch the product are high, the required profit margin must be sufficient to cover these costs as well as the manufacturing costs. Consequently, such firms typically include a life-cycle profitability analysis in their target costing systems. These systems use the target selling price and expected volume to determine the contribution the new product is expected to earn over its life. If this contribution is sufficient to cover the investment in the product plus make an adequate profit, then product development continues; otherwise the product is sent back for redesign.

Firms that are able to reduce the cost of their products substantially during the manufacturing stage of the product's life cycle undertake a different type of life-cycle analysis. For such firms, the selling price of the product presumably reflects the expected level of savings across the life of the product (in Japan, selling prices are rarely changed once the product is introduced). Therefore, the selling price is low compared to the launch manufacturing cost, and the initial profit is apparently too low. This problem is corrected by undertaking an analysis that includes the anticipated cost reductions in the life-cycle profitability estimates. Only if this analysis indicates that the product will make an adequate return is the product launched.

Once the future product mix of the firm has been established, individual product profit plans are established based on expected sales volumes. These individual plans identify the target selling prices and profit margins of the new products plus the overall sales revenues and profits generated. Summing these plans together provides a check to ensure that corporate sales and profit objectives will be met. Once the individual profit plans are approved, allowable costs are derived by subtracting the target profit margin for each new product from its target selling price.

The fourth question is designed to ensure that only achievable cost-reduction objectives are established. For target costing to be successful several conditions must apply. First, the target costs must be seen by everyone involved in the design process as the outcome of an unbiased process. Second, the cost-reduction objectives must be challenging but achievable most of the time. The Japanese set their target costs so that they create "tiptoe" objectives—that is, they can be achieved with considerable but not impossible effort. Third, the requirements for product functionality must be clearly understood.

Fairness is critical because everyone involved in the target costing process has to be committed to achieving the cost-reduction objectives. This commitment will be difficult to achieve if the objectives are viewed as the outcome of an imperfect analysis or a political process. The ability to achieve target costs is critical because most Japanese target costing systems rely heavily on the cardinal rule of target costing: the target cost of a product can never be exceeded. Without this rule, target costing systems lose their effectiveness. The primary objective of the cardinal rule is to stop the steady and relentless creep in product functionality and cost that occurs when product designers say, "If we just add this feature, the product will be so much better (and only cost a little more)." The required level of functionality (and quality) must be understood because the easiest way to remove costs from a product is to reduce its functionality. While some functionality reduction may be necessary to achieve the target costs, care has to be taken to ensure that product functionality does not fall below acceptable levels.

The fifth question requires the application of value engineering. VE is an organized effort to analyze the functions of goods and services so a firm can find ways to achieve those functions while meeting its target costs. VE helps to manage the trade-off between functionality and cost, the two dominant characteristics of the survival triplet. The objective of most Japanese VE programs is not to minimize the cost of products but to achieve a specified level of cost reduction established by the firm's target costing system.

The sixth question deals with special conditions in which the firm reduces the profit objective for the product even further and thus relaxes the target cost. Occasionally, firms find it necessary to launch products that exceed their target costs. These products appear to violate the cardinal rule, but often they have future revenues associated with them that are not captured in their individual selling prices. If these additional revenues are included in the target costing derivation, then launching these products can be shown to be beneficial to the firm.

The final question deals with decomposing the cost-reduction objective to the component level. Once the target cost of a new product is established, multifunctional product-design teams decompose it to determine component and subassembly target costs. The cost-reduction objective is not spread evenly across all of the major functions and the components and subassemblies they contain. Historical trends, competitive designs, and other data are used to estimate how much cost can be removed from each component or subassembly. An iterative process is used to ensure that once the component and subassembly target costs are set, they add up to the product's target cost. At many firms, the chief engineer is responsible for establishing the main themes of the new product. These themes are chosen to give the new product a distinctive character that the firm hopes will enable it to be successful when launched. For example, for an automobile the themes might be a quieter but sportier ride. The design team will allocate more costs to those features, but under the cardinal rule every extra dollar added to improve a feature of the vehicle must be taken out somewhere else.

Target costing is not a linear process that asks and answers each of the seven questions in sequence. Rather, it is a highly iterative process. Setting target costs often requires several cycles as the product's functionality, quality, and selling price are reassessed to enable the firm to meet its long-term profit objectives.

The Domain of Target Costing

While target costing systems focus primarily on direct costs, they also can be used to help reduce indirect costs. The big difference is that while direct costs can be affected at the individual product level, indirect costs are reduced only by changing the design of many products. For example, reducing the number of components in a single product might reduce direct costs but will not reduce the workload associated with parts management sufficiently to free up resources. If all new products have fewer components, however, then the overall reduction in parts complexity will be significant, and resources may be freed up.

Typically, Japanese companies do not have product cost systems that model indirect costs accurately. Instead of using activity-based costing systems, they still rely on traditional direct labor and machine-hour allocation procedures. However, this fact does not stop them from managing *future* indirect costs using a number of rules of thumb that can be viewed as faith drivers rather than proven cost drivers. These drivers include reducing the number of different materials used in a product, reducing the number of parts across the entire product line, and increasing parts commonality as much as possible across product lines without undermining the distinctive functionality of the individual models. Any low-volume components are flagged, actively designed out of existing products, and banned from future products.

Target costing identifies the costs at which future products must be manufactured if they are to earn the desired level of profits. Once the product-level target costs are set, then value engineering can be used to find ways to design the products so that they can be manufactured at their target costs.

VALUE ENGINEERING

Value engineering is *a systematic, interdisciplinary examination of factors affecting the cost of a product with the aim of devising a means to achieve its specified purpose at the required standards of quality and reliability and at an acceptable cost.* VE accomplishes this objective by analyzing the design of products to find effective ways to achieve their necessary functions and essential characteristics. The necessary functions define what the product must be able to do. The essential characteristics are the other requirements such as reliability, maintainability, and quality that must be satisfied if the product is to be successful.

An important aspect of Japanese VE programs is that their objective is not to *minimize* the cost of new products but to achieve a *specified* level of cost reduction. This distinction is important because designing to a target as opposed to minimum cost appears to lead to products with lower costs. The most likely explanation for this outcome is that designing to a specified low cost appears to create more intense pressure to reduce costs than designing to an unspecified minimum cost. This explanation is in keeping with Locke and White's research on goal setting in which they found that specific, challenging goals consistently lead to better performance than the general goal of doing one's best.[3] Later research by Locke and Latham also found that specific goals direct action more reliably than vague or general goals.[4]

---

[3] E. A. Locke and F. White, "Perceived Determinants of High and Low Productivity in Three Occupational Groups: A Critical Incident Study," *Journal of Management Studies,* 18: 1981, pp. 375-87.

[4] E. A. Locke and G. P. Latham, *Goal Setting: A Motivational Technique That Works!,* Englewood Cliffs, NJ: Prentice-Hall, 1984.

At the heart of VE lie two simple equations:[5]

$$\text{Value} = \text{function/cost} \tag{1}$$

and

$$\text{Perceived value} = \text{perceived benefits/price} \tag{2}$$

Equation (1) reflects the perspective of the producer, and equation (2) reflects that of the customer. Cost and price play the same internal and external roles in these equations as they do in the survival triplet. Consequently, VE practices are heavily integrated with consumer analysis and other techniques designed to ensure customer satisfaction. VE programs are as concerned with the product's final quality and functionality as they are with its cost.

The Five Key Questions of Value Engineering

The essence of the value engineering process can be captured in five questions identified by L. D. Miles, the founder of value engineering.[6] These questions are:

- What is it?
- What does it do?
- What does it cost?
- What else will do the job?
- What does that cost?

The first question deals with identifying the focus of analysis. When associated with a target costing program, the initial focus is the new product itself. However, as the value engineering process continues, the focus of analysis shifts to major functions and then components.

The second question deals with identifying the functions that the product is expected to perform. Function analysis is at the heart of VE. Two types of functions are identified, basic and secondary:[7]

- basic function: the principal reason for the existence of a "thing,"

---

[5] J. J. Kaufman, *Value Engineering for the Practitioner*, North Carolina State University, 1990, p. I-7.
[6] Ibid., p. I-10.
[7] Ibid., p. I-7.

• secondary function: a function that occurs because of the method selected to carry out the basic function, or those functions that support the basic function. Secondary functions can be wanted or unwanted.

To understand functions, they should be simply expressed using a verb and a noun. An incandescent light bulb "illuminates area." However, the incandescent light bulb also produces an unwanted secondary function, "generates heat," because of the method selected to achieve the basic function. Fire would also produce both functions, but "generates heat" may be the basic function in that situation.

Viewing a pencil in this way and identifying its components by functions produces something like this:[8]

Component	Function	Basic	Secondary
Pencil	Makes marks	*	
Eraser	Removes marks		*
Band	Secures eraser		*
	Improves appearance		*
	Transmits force		*
Body	Supports lead		*
	Transmits force		*
	Displays information		*
Paint	Protects wood		*
	Improves appearance		*
Lead	Makes marks	*	

Note that the lead in the pencil has the same function as the pencil itself, which is basic. All other functions, components, and costs to produce those functions are secondary. However, if the giver of the pencils determines that the pencil is an effective way to advertise his company, the basic function to the giver would not be the same as the user's basic function. Who then determines the value of a product? This may be a philosophical question, but in the final analysis it is still the user or the buyer, not the giver or producer, who determines the value of the products or services. The basic

---

[8] This example is adapted from Kaufman, 1990, op. cit., pp. II-9 through II-12.

function describes the intended use of the product, not how the product is being used. The pencil can also be used as a door stop, but that is not its intended use.

The third question in value engineering deals with determining the cost of the functions. Once the components have been identified, with their functions and costs, a component-cost-functions matrix can be created (see Table 4-2). Component cost is then allocated to the functions and totaled as to the cost and percent contribution to the overall product cost. The purpose of this step is to identify functions where value is low compared to cost. These functions are primary candidates for value engineering.

The value engineer pays specific attention to components that have high costs. For example, the function "remove marks" accounts for 40% of total costs and the functions "transmit force" and "make marks"s account for 10% and 17% respectively. Therefore, these three functions account for 67% of total cost. The value engineer would try to find ways either to reduce the cost of achieving the desired level of functionality or, if they are secondary functions, to eliminate them. For example, the eraser might be eliminated, thus reducing costs by 54% (i.e., the cost of the eraser and the band that secures it).

The fourth and fifth questions deal with finding alternative solutions to product design that provide increased value. This step can require great creativity on the part of the engineers. They must develop innovative solutions to the design problems they face. Cumulatively they must both increase the functionality of products and reduce costs. Individual design changes can affect either functionality or cost or both.

It is important not to take too narrow a view of achieving the target cost when undertaking VE. Superior functionality can sometimes lead to higher selling prices and hence higher target costs. However, the marketing group has to agree to a new target selling price before such cost increases can be authorized. Otherwise the cardinal rule of target costing will be violated. Alternatively, higher functionality at higher cost is acceptable if, and only if, the extra costs can be taken out of the product somewhere else without adversely affecting other forms of functionality or quality. If the additional costs can be offset, then product functionality can be increased without violating the cardinal rule. Otherwise, the

TABLE 4-2. THE COMPONENT-FUNCTION COST MATRIX OF A PENCIL

Function

Components	Cost cents	Remove marks %	Cost cents	Secure eraser %	Cost cents	Improve looks %	Cost cents	Make marks %	Cost cents	Transmit force %	Cost cents	Display info %	Cost cents	Support lead %	Cost cents	Protect wood %	Cost cents
Eraser	14.00	100.00	14.00														
Metal band	5.00			50.00	2.50	25.00	1.25			25.00	1.25						
Lead	6.00							100.00	6.00								
Body	5.00									50.00	2.50	10.00	0.50	40.00	2.00		
Paint	5.00					50.00	2.50									50.00	2.50
Total	35.00	40.00	14.00	7.10	2.50	10.60	3.75	17.00	6.00	10.60	3.75	1.50	0.50	6.00	2.00	7.10	2.50

superior functionality should not be introduced in this generation of the product.

SUMMARY

Target costing and value engineering are cost and profit management techniques that are applied during the design stage of a product's life. Focusing cost management on the design stage is critical because the ability to change the design of the product dramatically affects to what extent costs can be reduced. Experience has shown that it is easier to design costs out of products than to find ways to eliminate the costs after the products enter production. Thus, the great advantage of target costing and value engineering lies in their ability to act in a feedforward mode in the design stage as opposed to a feedback mode in the manufacturing stage.

At the heart of target costing lies a simple equation:

Target cost = target selling price – target profit margin

However, this simplicity hides a rich and complex practice developed by Japanese firms so they could become more competitive. With target costing they can design products that deliver the quality and functionality demanded by customers, at low cost.

The seven target costing and five value engineering key questions capture the logic behind target costing and value engineering. They do not capture the details of practice, which is complex and requires considerable expertise. While the target costing processes at firms vary considerably, they all share a common structure. This structure consists of three major parts, which are market-driven costing, product-level target costing, and component-level target costing. In the first part, market-driven costing is used to establish the allowable cost of each product (Chapter 5). In the second part, product-level target costing is used to set the target cost of each product (Chapter 6), and value engineering is used to find ways to achieve that cost (Chapter 7). In the final section, component-level target costing is used to set the target costs of the components (Chapter 8).

MARKET-DRIVEN COSTING

INTRODUCTION

The target costing process contains three major sections. It begins with market-driven costing, the purpose of which is to identify the allowable cost of future products. The allowable cost is the cost at which the product must be manufactured if it is to generate the desired profit margins when sold at its target price. In the next part, the product-level target costs are established. Finally, in the third part, the component-level target costs are established (see Figure 5-1).

The market-driven costing portion of the target costing process focuses on *customers* and their requirements and uses the allowable cost to transmit the competitive pressure in the market to the product designers and suppliers. Since the objective of target costing is to achieve an adequate profit margin across the life of the product, *life-cycle costing* is used to incorporate the up-front investment required to develop and launch the product and to compensate for any anticipated changes in the selling price of the product and its cost to manufacture over its life.

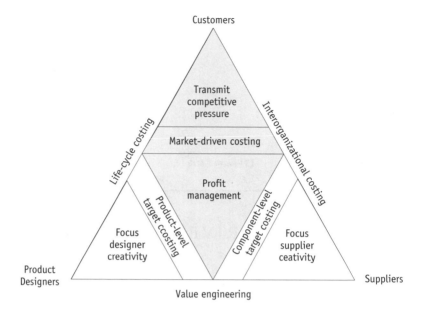

FIGURE 5-1. THE TARGET COSTING TRIANGLE:
THE MARKET-DRIVEN COSTING SECTION

Market-driven costing can be broken into five steps (see Figure 5-2). The first step, setting the long-term sales and profit objectives, highlights the primary role of target costing as a technique for profit management. The next step consists of structuring the firm's product lines to achieve maximum profitability. As part of this structuring the survival zones for future products are identified. Once a product's survival zone is located, the third step, setting its target selling price, can be undertaken.

Setting the target selling price is followed by the fourth step, which is establishing the target profit margin that the firm must earn to achieve its long-term profit objectives. Setting the target profit margin for products that need high capital investment or for which production costs are going to change significantly during the product's lifetime requires life-cycle product costing to ensure an adequate profit margin. Once this step has been completed, the allowable cost can be computed in the final step by simply subtracting the target profit margin from the target selling price.

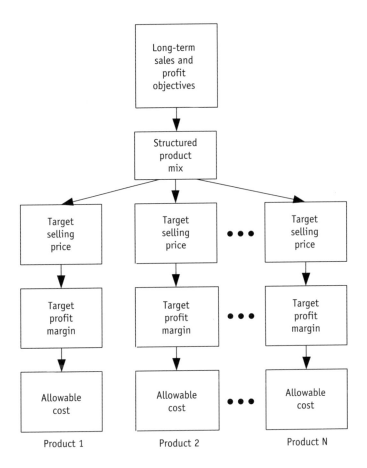

FIGURE 5-2. MARKET-DRIVEN COSTING

SETTING LONG-TERM SALES AND PROFIT OBJECTIVES

Target costing begins with the long-term sales and profit objectives of the firm. The primary objective of target costing is to ensure that each product over its life contributes its planned share of profits to the firm's long-term profit objectives. The credibility of the long-term plan is paramount in establishing the target costing discipline. Three factors contribute to that credibility: first, the plan is derived from careful analysis of all relevant information;

second, only realistic plans are approved; and third, the robustness of the plan is tested thoroughly.

To be credible, the plans have to be based on all relevant information about competitive conditions and other factors that influence product profitability. For example, at Olympus new products are introduced via the firm's extensive product-planning process. At the heart of this process is the product plan, which identifies the mix of cameras the firm expects to sell over the next five years. To ensure that the plan is practicable, the firm collects information from six sources: Olympus' corporate plan, a technology review, an analysis of the general business environment, quantitative information about camera sales, qualitative information about consumer trends, and an analysis of the competitive environment. It is the integration of these diverse sets of information that enables Olympus to set realistic product plans.

The planners at the firms take great care to ensure that the long-term plans are realistic and that wishful thinking is not driving the planning process. The entire target costing process is futile if the long-term profit objectives underlying it are unrealistic. The long-term profit plan usually is based on overall profit objectives that top management considers achievable. This constraint ensures that individual product profit objectives also will be achievable. Reflecting this objective, at Topcon the profit margin-setting process begins with the average profit margin for the entire business line, set by corporate strategy. For example, the medical and ophthalmic business unit is treated as a single business line, which is expected to generate a certain percentage profit each year. This percentage is derived primarily from past experience and is set to create a difficult but not impossible target for the unit.

The next test for realism is a firm's ability to achieve the sales volumes on which target costs will be based. At Toyota, the sales division proposes anticipated production volumes based on past sales levels, market trends, and competitors' product offerings. The sales division typically proposes a figure that is considered safe, that is, achievable. This figure is based on the model's current sales level. Optimism is restrained in favor of realistic goals.

Ensuring that the plan is practicable and that everyone involved in the product development process has bought into the plan requires an extensive review process. At Olympus, once the preliminary plan is completed, it is subjected to an exhaustive review to ensure that it is practical. The review covers issues such as the expected sales volume and profit for each camera model and the load such sales will place on the division's production and research and development resources. A team composed initially of research and development and product planning personnel conducts the review. Subsequently, as the product plan approaches acceptance, production personnel join the review team. Once the review is completed, a general meeting is held to accept the plan formally. This meeting is attended by division management, by the heads of the marketing, research and development, and production functions, and by the managers of the product-planning section. If they accept the product plan at this meeting, it is then implemented.

Finally, to test for robustness of the profit plan, the assumptions about how products generate profits are challenged. For example, some firms explore whether reductions in the historical profit margins will make it impossible for them to achieve their profit objectives. The point of this analysis is to determine if the proposed product mix is robust, that is, whether the profits would be adequate as long as the firm achieved its overall sales volume, irrespective of fluctuations in individual product sales volumes and profit margins. Nissan runs computer simulations to discover the answer. These simulations explore the impact on overall profitability of different price/margin curves for different product mixes. For example, historically higher margins have been earned on higher-priced vehicles, but there is no guarantee that this relationship will hold into the future. Therefore, some of the simulations explore scenarios with fundamentally different relationships between selling prices and margins.

The great care with which the long-term plans are developed illustrates the importance of their role in the target costing process. They create an umbrella under which the target costing system functions. Once these plans are established, the next step is to fine-tune the structuring of the firm's product line. Achieving the long-term profit plan is paramount. If it looks like overall the product line will not achieve its long-term profit objective, then actions are taken to increase overall group profitability. If these actions fail, the profit plan is amended. For example, at Sony if the planning process indicates that the group's target will not be met, the individual decisions made in the process of establishing the group's plan are reviewed. The planners consider three steps when this situation occurs. First, the engineers look for other ways to reduce the product's cost. Second, marketing is asked to raise prices and reduce the functionality of low-margin products to improve group profitability. Third, the general audio group and the marketing group will both accept reduced profitability if the product in question is considered critical to the corporation but cannot achieve its target profit margin.

STRUCTURING THE PRODUCT LINES

For product lines to be successful, they must be designed carefully to ensure that they satisfy as many customers as possible but do not contain so many products that they confuse customers. For that reason, structuring the product lines typically is based on a thorough analysis of the way customer preferences are changing over time. For example, at Nissan new models are conceptualized by identifying consumer mind-sets. Mind-sets capture characteristics of the way consumers view themselves in relation to their cars. These mind-sets can be used to identify design attributes that consumers take into account when purchasing a new car. Typical mind-sets include value seeker, confident and sophisticated, aggressive enthusiast, and budget/speed star. By detecting clusters of these mind-sets, Nissan can identify niches that contain a sufficient percentage of the automobile-purchasing public to warrant introducing a model tailored specifically for that niche.

To ensure that its product lines are well structured, Nissan develops a product matrix early in the design process that describes each vehicle by major market and body type (e.g., coupe or sedan). The information about each model in the matrix includes its price range, target customers and their income levels, and the range of body types supported. This information is maintained for both current and future models and effectively describes each model's market position. The primary purpose of the product matrix is to ensure that Nissan achieves the desired level of market coverage. New entries in the matrix are identified using consumer analysis. Market consulting firms undertake this analysis by using a number of different techniques, including general economic, psychological, and anthropological surveys, as well as direct observation.

Each product is expected to be profitable and so has to sell an adequate volume. To achieve this objective, a firm structures its product lines for maximum long-term profitability by making sure that the right number of products are in the line. For example, the range of models Sony produces is designed to strike a balance between profitability and customer service. Too many models would cause production and distribution costs to be too high. With too few models, too many sales would be lost to competitive products. The careful segmentation of the market shows in the relative sales volume of each model. With the exception of the professional models, all Walkman models sell in relatively high volumes. The highest-selling model accounts for about 10% of the total non-professional sales, and the lowest, for less than 1%.

The number of products in the line is subject to continuous analysis. For example, Nissan has recently decided to reduce the number of products it introduces. The logic behind this reduction is the increased cost of launching new vehicles. Olympus, in contrast, in the late 1980s increased the number of products in its line. The rationale behind this increase was to provide customers with cameras that better suited their life-styles. By increasing the number of products in its line, Olympus changed the nature of the way it competes using product functionality. The firm now differentiates its products horizontally on features such as waterproofness and size, and vertically by capability such as the zoom range of the lens. The horizontally differentiated products sell at the same price point, while the vertically differentiated products sell at different prices (see Figure 5-3).

This change in Olympus' strategy was prompted by the observation that market share associated with some price points was considerably larger than others. For high-volume price points, it was possible to identify different clusters of consumer preferences and profitably produce and market cameras designed specifically for those clusters. Under the new strategy, the number of models introduced at each price point is roughly proportional to the size of the market. Thus, the expected market share of each camera model offered is approximately the same unless it is designed to satisfy a low-volume, strategic price point.

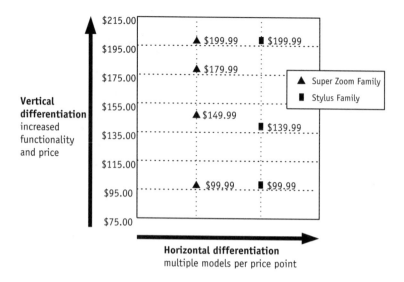

FIGURE 5-3. OLYMPUS OPTICAL'S VERTICAL AND HORIZONTAL DIFFERENTIATION STRATEGY

Once the product-line structure has been established, the survival zone of each product can be determined by consumer analysis. The survival zone defines the necessary functionality and quality a product must have to be successful and the price at which it will sell.

SETTING THE TARGET SELLING PRICE

The target costing process requires that a specific target selling price be established (see Figure 5-4). This price is critical in driving the target costing process. When firms sell the same product at different prices, for example in different countries, an average selling price is used. The target selling price has to be realistic, and consequently, the process of setting the target price is very thorough at most firms. Target selling prices are set by taking into account the market conditions expected when the product is launched. Nissan determines the target price by considering a number of internal and external factors. The internal factors include the position of the model in the product matrix and the strategic and profitability objectives of top management for that model. The external factors include the corporation's image and level of customer loyalty in the model's niche, the expected quality level and functionality of the model compared to competitive offerings, the model's expected market share, and the expected price of competitive models.

At the heart of the price-setting process is the concept of perceived value. Customers can be expected to pay more for a product than for its predecessor only if its perceived value is higher. For example, at Toyota the sales divisions usually propose retail prices and sales targets. The fundamental guiding principle used in setting the retail price is that the price remains the same unless there is a change in function from the previous to the new model and this change alters the perceived value of the vehicle in the eyes of the customer. Such a change might be the introduction of four-wheel steering and active suspension, or better performance from higher engine horsepower, or better fuel efficiency.

Many firms, including Toyota, take an incremental approach to the analysis. They start with the predecessor model's actual selling price and adjust it according to the incremental perceived value. The selling price of a new car model is viewed as being a

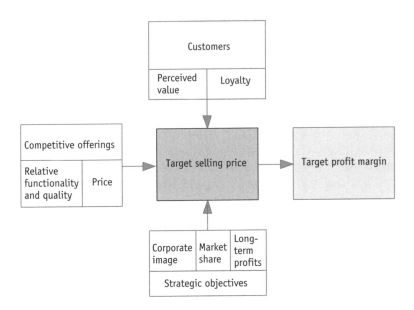

FIGURE 5-4. SETTING THE TARGET SELLING PRICE

combination of the selling price of the equivalent existing model plus any incremental value due to improved functionality. For example, adding air conditioning to the standard version of a model will increase its price by the value of air conditioning as perceived by customers. The incremental value of a new model is determined by analyzing market conditions. Due to the maturity of the automobile industry, most new features already exist in some form on other models. For example, if air conditioning is to be included in the standard version, its added value is determined using the list price of optional air conditioners for other models. If no equivalent option exists, a rare event, then the firm's design engineers and market specialists estimate how much customers are likely to be willing to pay for the added feature.

The interaction between functionality and price is carefully analyzed. Management takes the maximum allowable functionality into account when setting prices and deciding what functionality should be added. The aim is to design a product that will sell at its target price and achieve the desired sales volume. At Toyota, the price increase for an added function is not always equal to its selling price as a stand-alone option. The incremental price for a given increase in functionality might be lowered because of the firm's strategy for the vehicle model in question and because of the pricing strategies of competitors. As functions are added to the standard version, the selling price is raised until it reaches the upper limit for that class of vehicle, that is, the maximum selling price the firm believes it can set for the new vehicle. When this limit is reached, the only potential benefit from adding functionality is in increased sales. Thus, Toyota's strategy is to launch products that are in their survival zones, but whose functionality is as close to the highest feasible level as possible.

The price increases associated with incremental perceived value are tempered by the availability of competitive products and their perceived value. Selling prices can be raised only if the perceived value of the new product not only exceeds that of the product's predecessor but also that of competing products. For example, at Topcon competitive forces determine the allowable range of market prices. Topcon sets the price of its new products close to its competitors' prices. However, if management believes the Topcon product has greater functionality than competitive products, then the price of the Topcon product will be higher. If the functionality is perceived to be lower, the price will be correspondingly lower. Once the new product enters the market, competitors usually react by repricing their own products, increasing their advertising levels, or introducing a new model at a lower price.

The complexity of the task the firm faces in setting target selling prices depends primarily on the gap between product generations. It is particularly difficult when a firm launches a product that has no direct predecessor. For example, Citizen, a manufacturer of watches and watch movements, occasionally brings out a watch or movement for which there is no direct competitive offering. In such cases, the company determines the selling price using a "to be accepted" market price. It determines this price through market surveys and analyses that evaluate the attractiveness of the product and compare it with other watches and consumer products.

In contrast, when the new products are very similar to the ones being replaced, the task is simple. At Sony some Walkman models are introduced every year and the gap is small. Setting target selling prices is not particularly difficult because the new product is usually designed to replace the previous year's product and is quite similar. The sales volumes and selling prices of the previous product are good predictors of the new model's success.

In some industries there is very little price freedom for products. Therefore, the challenge facing the firms is simply to identify at which price point their products will sell. Sometimes the specific or distinctive functionality of the product determines this price point (see Figure 5-5). Olympus determines the appropriate price point for a camera by its distinctive feature, such as magnification capability of the camera's zoom lens or the camera's small size. The competitive analysis and technology review during development of the product plan determine the relationship between distinctive features and price points. The product plan thus describes cameras only in terms of their distinctive features.

Other features are added as the camera design nears completion. In some markets such as cameras, the price/functionality trade-off means that over time the price point for a given level of functionality falls. For such firms, the challenge is to introduce products with new levels of functionality to maintain the high price points while

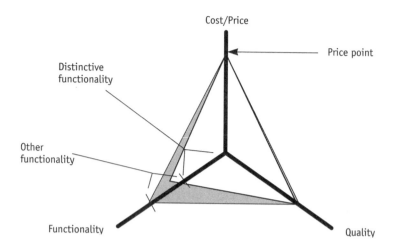

FIGURE 5-5. THE SURVIVAL ZONE OF A CAMERA

simultaneously decreasing the costs of existing levels of functionality sufficiently to generate adequate profits as prices fall (see Figure 5-6). At Olympus the price point at which a camera with given functionality sells tends to decrease over time with improvements in technology. To maintain the price points for as long as possible the functionality of the camera is enhanced periodically, for example, by adding a quartz date/day feature or panoramic capability. However, as the cost of the underlying distinctive functionality drops, a point is reached at which minor functionality enhancements can no longer support the original price point, and the price is reduced to the next lower price point. The natural outcome of this process is to generate new price points at the low end. For example, the price point for the simplest compact camera was $150 in 1987 and $100 in 1990.

If functionality can be increased quickly enough, products at new higher price points can be introduced. At Olympus the high-end technology generates new price points. As the functional gap between the capabilities of compact and SLR cameras closes, it becomes possible to introduce compact cameras at higher prices. For example, Olympus created a new price point of $300 when it introduced the first compact camera with 3X zoom capability in 1988. Over time, the number of price points had increased.

Target selling prices are established within the context of the firm's long-term sales and profit objectives. The target selling price must take into account the market share the firm wants the product to achieve, its profitability objectives, and the image the firm is trying to project via its products. Obviously, the lower the selling price, the higher the market share, but decreasing the selling price below a certain point will lead to reduced profitability even though sales are higher. Therefore, the selling price also has to accommodate the long-term profit objectives of the firm.

Finally, the image the firm wants to project via its products has to be considered. If the firm is trying to establish an image of its products as being high value, then the selling price might be lowered to strengthen this image. In contrast, if the image is one of technological superiority, then the price might be increased. Thus, target selling prices reflect a trade-off among these three strategic objectives.

Setting target selling prices is a complex undertaking requiring careful analysis of how the customer perceives value and how competitive offerings deliver value. Given the importance of the target selling price to the target costing process, it is not surprising that firms are very careful to ensure the most realistic possible target selling prices.

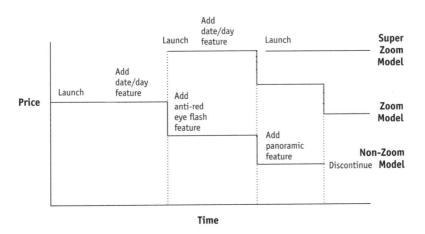

FIGURE 5-6. PRICE AND FUNCTIONALITY FOR A CAMERA OVER TIME

Setting the Target Profit Margin

The objective in setting target profit margins is to ensure achievement of the firm's long-term profit plan (see Figure 5-7). Usually the division in charge of the product lines is responsible for achieving the overall profit target. For example, at Sony an iterative process sets the target profit margin for each product. The starting point is the group profit margin as identified by the group profit plan. Sony's corporate planning department develops this plan after negotiations with the planning department of the general audio group. Once the annual group profit target is set, the group is responsible for its own profitability.

The two considerations in setting target profit margins are to ensure that they are realistic and that the margin is sufficient to offset the life-cycle costs of the product.

Setting Realistic Target Profit Margins

Firms set target profit margins in two ways. The first starts with the actual profit margin of the predecessor product and then adjusts for changes in market conditions. Nissan adopts this approach. It uses a computer simulation to identify the relationship between

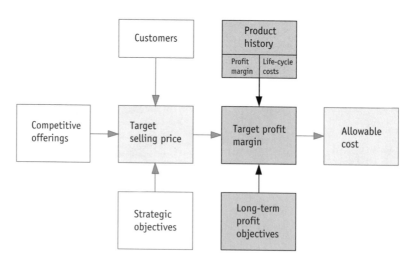

FIGURE 5-7. SETTING THE TARGET PROFIT MARGIN

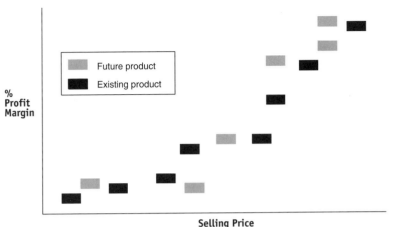

Source: Harvard Business School case 194-040. Adapted and reprinted by permission.

FIGURE 5-8. NISSAN MOTOR COMPANY, LTD.: IDENTIFYING THE TARGET PROFIT MARGIN

selling prices and profits and from this relationship identifies the target profit margins of future products, based predominantly on their target selling prices (see Figure 5-8). Each new model's target profit margin is established by running simulations of the firm's overall profitability for the next 10 years, assuming it is selling the models identified in the product matrix at their expected sales volumes. The simulations start by plotting the actual profit margins of existing products against their selling prices. The desired profitability of planned models at their target selling prices is then added, and the firm's overall profitability is determined over the years at various sales levels. This predicted overall profitability is compared to the firm's long-term profitability objectives set by senior management. Once a satisfactory future product matrix is established that achieves the firm's profit objective, the target profit margins for each new model are set. The aim of this careful analysis is to set realistic profit margins that will enable the firm to achieve its long-term profit plans.

The second method for setting target profit margins starts with the target profit margin of the product line (or other grouping of products) and raises or lowers the target profit margin for individual products depending on the realities of the marketplace. To illustrate this approach, Sony starts with the target profit margin for the entire product line. It calculates the first-cut target cost for a new product by subtracting the group target profit margin from the product's target selling price. It compares the resulting target cost to the estimated cost of the new product. When the target cost is considered too low, the target profit margin is allowed to decrease if another product with a sufficiently high target profit margin can be identified to offset the loss. Thus, if one product's profit margin has to be decreased, another's profit margin has to be increased. When all the individual product decisions are complete, the firm runs a simulation of overall group profitability to make sure the group's target will be met.

Thus, the target profit margin is set based on historical profit levels, the relative strength of competitive offerings, and the profit objective captured in the long-term profit plan. If management feels this profit objective is unrealistic, it is modified and target profit margins are reduced, increasing the allowable costs. Setting target profit margins in this manner makes the allowable costs reflect the relative competitive position of the firm. A highly efficient firm will set target profit margins higher than less efficient firms and will have lower allowable costs.

Adjusting the Target Profit Margin for Life-Cycle Costs

If product launches or discontinuances require high investments or if their selling prices and costs are expected to change significantly during their lives, the target profit margin has to be adjusted accordingly. The purpose of these adjustments is to make sure that the expected profitability of the product across its life is adequate. Firms with products that require large up-front investments carry out life-cycle analyses so they can set target profit margins large enough to ensure that the products earn an adequate profit margin over their life.

For example, at Nissan when the conceptual design reaches the stage where rough estimates of the number of vehicles likely to be sold and the costs associated with development of the new products can be made, the firm undertakes a life-cycle contribution study to estimate the overall profitability of the proposed model. The purpose of this study is to ensure that the new model is likely to generate a positive contribution over its life. The study compares the estimated revenues generated by the new model to the expected cost of the product across its life.

The life-cycle analysis sometimes requires modification of the cost reported by the firm's financial accounting system, to give a more accurate picture of the capital costs associated with the new product. At Nissan, the depreciation charge used for the life-cycle contribution analysis is not the one used for financial reporting purposes. Nissan reports depreciation using a declining balance method for both tax and financial reporting purposes. However, for the life-cycle contribution calculation it uses a straight-line approach. Management modified the depreciation calculation because it felt that the straight-line approach better captures the relationship between asset use and models produced than the declining balance approach.

If the life-cycle contribution is deemed satisfactory, the conceptual design process is allowed to continue. If the life-cycle analysis shows an unsatisfactory contribution, the product is subjected to redesign. When the product development cycle is long, multiple life-cycle analyses are sometimes undertaken. Typically, such life-cycle analyses are performed at each major design step to ensure that the product will support the firm's profit objectives. Nissan conducts a major review of the new model after completing the first value engineering stage. This review includes an updated profitability study and an analysis of the performance characteristics of the model. The profitability study compares the expected profitability of the model given by the target price minus the target cost to the latest estimates of the capital investment and remaining research and development expenditures required to complete the design of the product and allow production to commence.

Firms that are able to reduce the cost of their products substantially during the manufacturing stage of the product's life cycle undertake a different form of life-cycle analysis. The purpose of this analysis is to reflect any savings in production costs made during the manufacturing phase in the target costing profitability analysis. For example, the joint impact of increased functionality and reduced prices places severe pressure on Olympus to reduce cost. Only by continuously taking costs out of products can the firm hope to remain profitable. This intense cost reduction across the manufacturing life of the product often means that the target selling price already reflects anticipated cost savings. The target cost is set for some future point during the product's manufacturing life, not the first few months of production. If it is not possible to increase the price or to reduce the production cost sufficiently to reduce the estimated cost below the target cost, then a life-cycle profitability analysis is performed. This analysis includes the effect of potential cost reductions over the production life of the product in the financial analysis of the product's profitability. Olympus expects to reduce production costs by about 35% across the production lifetime of its products. The product is released if these life-cycle savings are sufficient to make the product's overall profitability acceptable.

Life-cycle adjustments help ensure that all costs and savings are taken into account when determining the target profit margin. Without such adjustments, the firm risks launching products that will not earn an adequate return over their lives. Given that these life-cycle costs are taken into account in the target profit margin, the computed target cost includes only manufacturing costs. Usually at the end of the development cycle the target cost reflects all manufacturing costs, both direct and indirect.

SETTING THE ALLOWABLE COST

Once the target selling price and target profit margin have been established, the allowable cost can be calculated (see Figure 5-9):

Allowable cost = target selling price − target profit margin

Given the way target profit margins are set, the allowable cost reflects the relative competitive position of the firm. In confrontational environments, firms that are highly efficient will have higher target profit margins and hence lower allowable costs than their less efficient competitors. Consequently, allowable costs are not benchmarks against which the firm can measure itself compared to its competitors. To make allowable costs act as such external benchmarks, target profit margins must be set that reflect the capabilities of the most efficient competitor. Such margins sum to give benchmark profits, not the lower, realistic, long-term profit objectives of the firm.

The Japanese firms in the sample have chosen to set achievable profit margins and thus have the target costing system articulate with the long-term profit objectives of the firm. Unfortunately, this practice obscures the cost implications of the relative competitive position of the firm. This position can be highlighted by estimating benchmark costs and comparing them to the firm's allowable costs.

Firms that are at a significant competitive disadvantage will benefit most from estimating benchmark costs and calculating the difference between those costs and their allowable costs. If the disadvantage is significant, it might not be possible to reach the benchmark costs in a single generation of product design. Such firms will have to adopt a multigenerational strategy of product design, setting ever more aggressive cost targets for each generation. The narrowing gap between the benchmark and allowable costs would demonstrate the achievement of competitive parity. Firms that are highly effective will not benefit from benchmarking and may even be hurt by it if they become complacent. For such firms the calculation of waste-free costs described in the next chapter is more appropriate.

Thus, the allowable cost represents the cost at which, according to top management, the product must be manufactured if it is to achieve the target profit margin when sold at its target price. It acts as a signal to all involved in the target costing process as to the magnitude of the cost-reduction objective that eventually must be achieved.

FIGURE 5-9. SETTING THE ALLOWABLE COST

SUMMARY

Target costing is as much a technique for profit management as it is for cost management. The process begins with the establishment of the long-term sales and profit plans of the firm. These plans are based on a careful segmentation of the market so that a limited number of products satisfy the maximum number of customers. Once the products in the product line have been identified, their target selling prices can be established.

The location of the survival zone of a given product determines its target selling price. The process of setting target selling prices includes a careful analysis of competitive conditions and customer expectations. Once the target selling price is set, a target profit margin is established, based on the firm's experience with similar products (including previous generations) and on evaluation of competitive offerings. The target profit margin must also take into account life-cycle effects. For example, if the product requires high up-front investments, then the profit margin will have to be high enough to recover these costs over the life of the product. Alternatively, if the manufacturing cost of a product can be decreased over its production life the initial profit margin can be reduced accordingly.

Subtracting the target profit margin from the target selling price gives the allowable cost (see Figure 5-10). The allowable cost identifies the cost at which the product must be manufactured if it is to generate its target profit margin. The allowable cost does not reflect the capabilities of the firm or its suppliers and therefore are often unachievable. When allowable costs are not achievable, the target cost of the product is set higher than the allowable cost. Setting achievable product-level target costs is the subject of the next chapter.

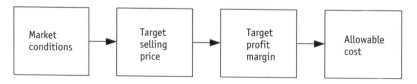

FIGURE 5-10. CALCULATING THE ALLOWABLE COST

CHAPTER 6

PRODUCT-LEVEL TARGET COSTING

INTRODUCTION

Once the first section of the target costing process, market-driven costing, is complete the second section can commence. This section consists of setting achievable, product-level target costs. These costs are equal to or higher than the allowable costs set in the price-driven section of the process. The second section of the target costing process focuses the creativity of the product designers on finding ways to design products that satisfy the firm's customers at the allowable cost. It uses value engineering to achieve this objective and communicate it to the firm's suppliers.

In practice, however, it is not always possible for the designers to find ways to achieve the allowable cost and still satisfy the customers. The process of *product-level target costing* increases the allowable cost of the product to a level that can reasonably be expected to be achievable, given the capabilities of the firm and its suppliers (see Figure 6-1).

Product-level target costing can be broken into three steps (see Figure 6-2):

- setting the product-level target cost;
- disciplining the target costing process by monitoring

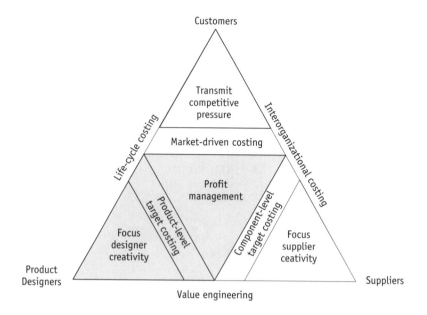

FIGURE 6-1. THE TARGET COSTING TRIANGLE:
THE PRODUCT-LEVEL TARGET COSTING SECTION

progress, applying the cardinal rule of target costing, and allowing for extenuating circumstances;

- using value engineering and other techniques to achieve the product-level target cost.

Value engineering is described in the next chapter.

SETTING THE PRODUCT-LEVEL TARGET COST

In highly competitive markets, customers expect each generation of products to improve. These improvements can be improved quality or functionality or reduced prices. Any of these improvements or a combination require that the firm reduce costs to maintain its profitability. The degree of cost reduction required to achieve the allowable cost is called the cost-reduction objective. It is derived by subtracting the allowable cost from the current product cost:

Cost-reduction objective = current cost − allowable cost

The current cost of the new product is determined by summing the current manufacturing cost of each major function of the new model. No cost-reduction activities are assumed in computing the current cost of the product. For the current cost to be meaningful, the major functions used in its construction have to be very similar to those that eventually will be used in the new product. For example, if the existing model uses a 1.8 liter engine and the new model uses a 2.0 liter one, the appropriate current cost is for the most similar 2.0 liter engine the firm produces or purchases.

Since the allowable cost is derived from external conditions and does not take into account the internal design and production capabilities of the firm, the risk is that the allowable cost will not be achievable. In this case, to maintain the discipline of target costing the firm has to identify the achievable and the unachievable part of the cost-reduction objective. The achievable or *target cost-reduction objective* is derived by analyzing the ability of the product designers and suppliers to remove costs from the product (see Figure 6-3). The process by which costs are removed from the product is called

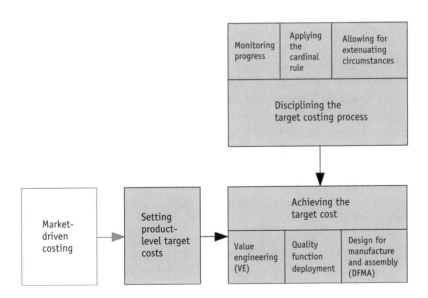

FIGURE 6-2. PRODUCT-LEVEL TARGET COSTING

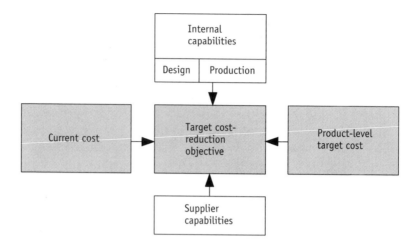

FIGURE 6-3. SETTING THE TARGET COST-REDUCTION OBJECTIVE

value engineering. It creates an interactive relationship with the suppliers. The purpose of this relationship is to allow the suppliers to provide early estimates of the selling prices of their products and, when possible, insights into alternative design possibilities that would enable the firm to deliver the desired level of functionality and quality at reduced cost.

The unachievable part of the cost-reduction objective is called the *strategic cost-reduction challenge* (see Figure 6-4). It identifies the profit shortfall that will occur because the designers are unable to achieve the allowable cost and signals that the firm is not as efficient as demanded by competitive conditions. Typically, in a firm with a well-established and mature target costing system, the strategic cost-reduction challenge will be small or nonexistent, and intense pressure will be brought on the design team to reduce it to zero.

For the most efficient firms, the achievable cost reduction for a product might exceed the cost-reduction objective. Such firms do

FIGURE 6-4. THE STRATEGIC COST-REDUCTION CHALLENGE

not face a strategic cost-reduction challenge. Their superior efficiency lets them increase market share by reducing the selling price
of the product or by increasing its functionality while keeping the
price constant, or it lets them earn higher profits by keeping both
the price and functionality at their targeted levels. In confrontational
environments such conditions are short lived because firms can achieve
only temporary competitive advantage over their competitors.

For the discipline of target costing to be maintained, the size of
the strategic cost-reduction challenge must be managed carefully. A
strategic cost-reduction challenge should reflect the true inability of
the firm to match its competitors' efficiency. To ensure that the strategic cost-reduction challenge meets this requirement, the target cost-
reduction objective must be set so that it is achievable only if the
entire organization makes a significant effort to reach it. If the
target cost-reduction objective is consistently set too high, not only
will the work force be subjected to excessive cost-reduction objectives, risking burn-out, but also the discipline of target costing will
be lost as target costs are exceeded frequently. If the target cost-
reduction objective is set too low, the firm will lose competitiveness
because new products will have excessively high target costs.

The product-level target cost is determined by subtracting the
new product's target cost-reduction objective from its current cost
(see Figure 6-5). That is:

Product-level target cost = current cost – target cost-reduction objective

The strategic cost-reduction challenge is determined by subtracting
the allowable cost from the target cost (see Figure 6-6):

Strategic cost-reduction challenge = target cost – allowable cost

The value of differentiating between the allowable cost and the
product-level target cost in this manner lies in the discipline that it
creates. In many firms the allowable cost is too low to be achievable, given the capabilities of the firm and its suppliers. Target
costing systems derive their strength from the application of the
cardinal rule: the target cost must never be exceeded. If a firm continuously sets over-aggressive target costs, violations of the cardinal rule would be common and the discipline of the target costing
process would be lost. Even worse, if the allowable cost is known
to be unachievable, the design team might give up even trying to

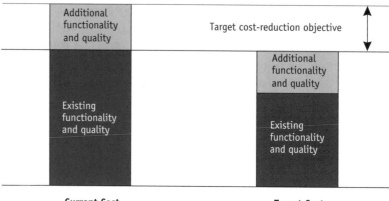

FIGURE 6-5. THE TARGET COST-REDUCTION OBJECTIVE

achieve it, and cost reduction would cease to occur. To avoid this motivation problem, firms frequently set target costs higher than the allowable costs. These target costs are designed to be achievable but only with considerable effort. They allow the cardinal rule to be maintained for almost every product.

The distinction between allowable and product-level target costs thus plays two roles. First, it identifies the strategic cost-reduction challenge, which creates a powerful pressure on the design team of

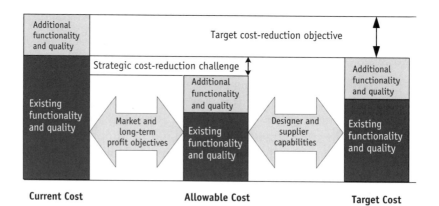

FIGURE 6-6.   DETERMINING THE STRATEGIC COST-REDUCTION
             CHALLENGE

the next generation of the product to be even more aggressive about cost reduction. In this way, the failure to achieve the allowable cost this time around is turned into a challenge for the future, not a current defeat. Second, defining an allowable cost avoids weakening the cardinal rule, which applies only to target costs, not allowable costs. The process by which the strategic cost-reduction challenge is established must be highly disciplined. Otherwise it becomes a mechanism to reduce the effectiveness of target costing by setting target costs that are too easy to achieve. In most firms, senior management approves the strategic cost-reduction challenge before the product-level target cost can be set.

The setting of the strategic cost-reduction challenge at the firms in the sample is based on an allowable cost that in turn is based on a target profit margin that is considered achievable. The problem with this approach is that the strategic cost-reduction challenge captures only a portion of the firm's relative efficiency compared to its competitor. Switching to a benchmark allowable cost that sets the target profit margin at the level of the most efficient firm in the market allows all the relative competitive disadvantage to be captured. The strategic cost-reduction challenge now shows how far behind the firm is compared to its most efficient competitor.

The problem with competitive benchmarking is that it can engender complacency in the most efficient producers. Such complacency is a recipe for disaster when lean enterprises are competing because the other firms will rapidly overtake the complacent firm. Lean enterprises, with their continuous drive to reduce waste, should always benchmark themselves against a waste-free standard and not just against the best competitor.

Waste-Free Target Costs

For benchmarking purposes, the product-level target cost should be compared to the *waste-free cost* of the product. There are two general types of waste. The first type does not create value for the customer but is necessary for the firm to achieve its objectives and therefore cannot be eliminated immediately. For example, inventories required to ensure a reliable source of components are necessary but do not add any value to the product. Eliminating such inventories without finding new ways to maintain a steady supply of components would lead to customer dissatisfaction due to delivery delays. Therefore, such waste is not avoidable in the immediate future.

The second type of waste neither creates value nor is necessary and should be eliminated as soon as possible. For example, a process that has a high defect rate and causes the downstream processes to wait creates waste. If the underlying causes of the defects are correctable, given existing technology, then the waste is avoidable.

Two waste-free costs can be computed. The first assumes that no nonvalue-added activities are performed and that all value-added activities are performed as efficiently as possible. This cost is the *perfect waste-free cost*. The lean enterprise's ultimate long-term goal is to try to achieve this cost. It can be likened to the zero-defects objective in quality management. The second waste-free cost assumes that only necessary nonvalue-added activities are performed and that all value-added activities are performed as efficiently as possible. This cost is the *unavoidable waste-free cost*; it is the most aggressive short-term cost-reduction goal possible for the product. The difference between the product-level target cost and the two waste-free costs identifies how much waste the firm is currently accepting in the product.

In most, if not all, cases, the perfect waste-free cost will be below the allowable cost. It is not meant to be achieved but creates the perfection objective toward which everyone must strive. It ensures that complacency will not cause even the most efficient firms to slow in their relentless efforts to reduce costs. The value of absolute cost-reduction objectives is that firms can see how far they have to go in becoming truly lean. In particular, these objectives help the most efficient firms maintain an intense pressure to reduce costs. The unavoidable waste-free cost can be above or below the allowable cost, but usually it will be below the allowable cost, indicating that the firm still has some unavoidable waste in its processes. If the unavoidable waste-free cost is higher than the allowable cost, then the firm must face some structural cost impediment that causes its costs to be systematically higher than its competitors' costs.

The calculation of the waste-free costs can thus augment the target costing process by placing the current cost and product-level target costs in the perspective of the cost-reduction path to the ultimate waste-free objective (see Figure 6-7). Achieving this objective

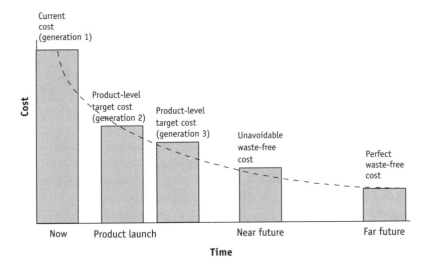

FIGURE 6-7. THE COST-REDUCTION PATH

starts at the point where the firm is today (the product's current cost). It continues to where the firm expects to be in the near future (the product-level target cost) and to where it should be within a few generations (the unavoidable waste-free cost[1]) and ends where it eventually should be (the perfect waste-free cost[2]). The allowable cost thus identifies the cost-reduction demands of the market, the product-level target cost identifies the firm's current capabilities, and the waste-free cost shows how much farther along the cost-reduction path the firm still has to go.

The Transition from Allowable to Product-Level Target Cost

Technically, the target cost of a product is the target selling price less the target profit margin plus the strategic cost-reduction

---

[1] The unavoidable waste-free cost will decrease over time as more ways to avoid currently necessary nonvalue-added activities are identified. Here, the unavoidable waste-free cost refers to the currently identified cost.

[2] Theoretically, the perfect waste-free cost of a given product is constant over time. However, as the functionality of the product changes with each generation, so must its perfect waste-free cost.

challenge. However, many firms blur the distinction between the allowable cost and the target cost by stating that the target cost is determined by subtracting the target profit margin from the target selling price. This simplification makes it easier for people to understand the spirit of target costing as being price driven. Obviously if the strategic cost-reduction challenge is zero, the allowable and target costs are identical.

At some firms, even when the allowable cost is considered achievable, it is not referred to as a target cost until the process has reached the stage at which the major component target costs are established. The retention of the term "allowable costs" shows that top management is not willing to invoke the cardinal rule until they are convinced that the target cost is indeed achievable. For Nissan this point is not reached until the groups responsible for the major functions have signed off on their individual target costs.

For complex products with a long design cycle, intermediate partial target costs are often generated. For example, once Nissan has developed the target costs for its major functions, it establishes the draft target cost for the new product. Later in the design process, when the assembly and indirect manufacturing costs are established, these costs are added to the draft target cost to give the final target cost. Thus, the product-level target cost consists of three elements: major function-level target costs, assembly target costs, and indirect manufacturing target costs. The assembly and manufacturing costs typically are managed using either design for manufacture and assembly and/or quality function deployment techniques. The discipline of target costing is maintained by not allowing the partial targets to change over the life of the project. For products such as Olympus cameras that have much shorter life cycles, intermediate target costs usually are not established.

Adjusting the Target Cost for Changes in Market Conditions

When market conditions change, firms may have to revise their target costs. Such revisions have to be managed carefully. For example, at Olympus the target costing system is designed to identify when the anticipated selling price has changed sufficiently to render the original calculations of target costs obsolete. This objective is achieved through the use of cost ratios determined by dividing the product-level target cost by the free-on-board (FOB) target price. The target cost ratio for a given camera is set based on the historical cost ratios of similar cameras, the anticipated relative strength of competitive products, and the overall market conditions anticipated when the product is launched. Once the target cost ratio is established, it is converted into yen by multiplying it by the target price. This yen-denominated target cost is used in all future comparisons with the estimated cost of production to ensure that the appropriate level of cost reduction is being achieved.

During the design phase, the anticipated cost ratio (that is, what the firm actually expects to happen) of new products is monitored on a frequent basis. The FOB price of a new product is sensitive to both market conditions and fluctuations in foreign exchange rates. Olympus sells 70% of its cameras overseas, and the FOB price of a product is the weighted average yen price. Since the FOB price for cameras sold overseas is designated in the appropriate foreign currency, fluctuations in the exchange rates cause the yen-denominated FOB price to change.

If the FOB price changes sufficiently during the design phase to cause the anticipated cost ratio for the camera to fall outside the acceptable range, then the target cost of the camera is reviewed and typically revised to bring the anticipated cost ratio back into the acceptable range. If the FOB price is falling, as has been the case for most of the 1990s, the result is a lower target cost, which makes it harder to achieve. If it is rising, the result is a higher target cost, which allows the functionality of the product to be increased or higher profit margins to be earned.

Thus, the target cost is sometimes modified to reflect changes in market conditions. The primary objective of target costing is to achieve the profit margins assumed by the firm's long-term profit plan, and so some degree of updating is necessary. However, such modifications to target costs are carefully managed; otherwise the original target cost loses its meaning and the discipline of the cardinal rule will be lost.

At Olympus, the term "target cost" is used only after an achievable cost-reduction objective is decided. Like Nissan, they have developed a procedure to ensure that only achievable target costs are set (see Figure 6-8). At Olympus, the target cost is based on the price point for the distinctive feature of the camera. Research and development is responsible for identifying the other features of the camera (e.g., the type of flash and shutter units). Feature identification is an iterative process that estimates the cost of each new design and compares it to the product's target cost.

Approximately 20% of the time the estimated cost is equal to or less than the target cost, and the product design is released for further analysis by the production group at the Tatsuno plant. The other 80% of the time the research and development group must

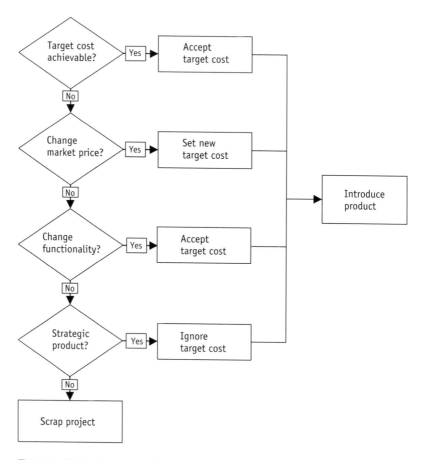

FIGURE 6-8. OLYMPUS OPTICAL COMPANY LTD.:
INTRODUCING THE PRODUCT

do further analysis. First, they ask marketing if the price point can be increased sufficiently to make the target cost equal to the estimated cost. If the price can be increased sufficiently, the product is released to the production group. If the market price cannot be increased sufficiently, then research and development explores the effect of reducing the functionality of the product. Reducing the product's functionality decreases its estimated cost to produce. Only if these reductions are sufficient is the product released to production. Otherwise the project is scrapped or returned to R&D for complete redesign unless the product is considered strategic.

Simply setting a product-level target cost is necessary but insufficient for target costing to be effective. The next challenge is to create a disciplined environment that enables the target cost to be achieved.

DISCIPLINING THE PRODUCT-LEVEL TARGET COSTING PROCESS

Disciplining the product-level target costing process begins with monitoring the progress of the design engineers toward reaching the cost-reduction objective. It continues with the application of the cardinal rule of target costing. Sometimes, the cardinal rule has to be applied in a more sophisticated way than the conventional, single-product perspective. When one product leads to increased sales of other products, a multiproduct perspective has to be adopted; when it will lead to sales of future generations of products, a multigenerational perspective is required. It is only when getting the product to market is so imperative that cost is of secondary consideration that the cardinal rule may be violated. Finally, when the product is released for mass production and its actual cost of manufacturing can be measured, steps sometimes have to be taken to reduce those costs to the target level.

Monitoring Progress Toward Achieving the Target Cost-Reduction Objective

Once the target cost-reduction objective has been established, the process of designing the product so it can be manufactured at its target cost can commence. The discipline of target costing requires that the chief engineer and his superiors continuously monitor the progress the design engineers are making toward this objective. This monitoring ensures that corrective actions can be taken as early as possible and that the cardinal rule will not be broken.

Some firms define an *as-if cost* at this point in the development process. The as-if cost reflects cost-reduction opportunities identified when the previous generation of the product was being designed or manufactured. In most cases the as-if cost is above the target cost of the new product. Therefore, the additional cost reduction that has to be achieved is given by the difference between the target cost and the as-if cost.

As the design process proceeds and costs are removed from the major functions, the estimated manufacturing cost gradually falls toward the target cost. Many firms call the updated estimate the *drifting cost* (see Figure 6-9). Thus the product design process starts with a current cost that is higher than the target cost and that across the design process reduces the expected or drifting cost until it finally reaches the target cost. At most firms, once the drifting cost equals the target cost, cost-reduction activities cease. There is no reward for achieving greater savings than those required to achieve the target cost. The engineers' time would be better spent

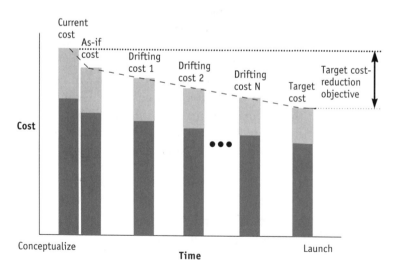

FIGURE 6-9. ACHIEVING THE TARGET COST

on getting the drifting cost of other products to equal their target costs.

The process of comparing the drifting cost to the target cost continues throughout the design process. For example, at Olympus, when the product is ready to be released to production, planners undertake a final review of the feasibility of the target cost. If the estimated production cost is too high, the design is subjected to additional analysis. Frequently, relatively minor changes in the product's design are all that is needed to reduce the cost estimate to the target cost level. As long as these changes do not change the product's price point, then the functionality is changed and the product is submitted for approval. If the design changes will reduce the price point, the product is returned to the research and development group for redesign.

For products that feature a variety of options, the final fine-tuning of the target cost is often achieved by specifying the features the standard product will contain. For example, if the manufacturing cost is too high, one or more "standard" features might be converted to options that the customer has to pay for. Converting features to options both reduces the cost of manufacturing the standard product, allowing the target cost to be achieved, and increases the selling price of the originally specified product, allowing the target profit to be achieved. The reduction in the functionality of the standard variant has to be subjected to market analysis to ensure that it is acceptable in the eyes of the customer. For example, the passenger-side air bag might be converted from a standard feature to an optional one. This reduction in functionality will be acceptable as long as competitive offerings treat the passenger-side air bag in the same way. This fine-tuning process allows firms more leeway to achieve target costs set several years before. For example, at Toyota the exact functionality of the standard version is set only when factors such as competitive offerings, foreign exchange rates, and user demand are better understood. Changing the functionality of the standard version allows Toyota to increase the probability that the new model will achieve its desired level of profitability.

Similarly, the actual selling price is not fixed until just before the product is launched. Delaying these two critical decisions significantly increases the probability that the firm will achieve the target cost. For example, the incremental value assigned to an air bag in the U.S. market might be $450, but the competition sets the incremental value at $700. In this case, Toyota might increase its price by the difference. Alternatively, if the competitive prices are lower, Toyota will drop its prices to match.

Thus, the key to success of the target costing process is getting the drifting cost equal to the target cost. If the cost-reduction objective cannot be achieved, then the product must either be more fundamentally redesigned to reduce its cost or abandoned. For example, at Sony the estimated cost of the new product is compared continuously to its target cost. If it appears that the new product will fail to meet the firm's minimum profit margin requirement or will adversely affect the ability of the group to meet its profit target, the product is redesigned.

Applying the Cardinal Rule

The cardinal rule of target costing, the target cost must never be exceeded, plays a critical role in ensuring that the discipline of target costing is maintained throughout the design process. The cardinal rule is enforced in three ways.

- Whenever improvements in the design result in increased costs, alternative, offsetting savings have to be found elsewhere.
- Launching products whose costs exceed their target is not allowed.
- The transition to manufacturing is managed carefully to ensure that the target cost is indeed achieved.

Finding Offsetting Savings

Application of the cardinal rule of target costing is illustrated by a design analysis example at Komatsu. The objective of redesigning the engine, transmission, and torque converter was to decrease the service time required—an enhancement to the functionality of the product. Under target costing, an increase in functionality or quality typically is not allowed to be offset by an increase in cost. So when a new approach costs more, the design team is forced to look elsewhere for additional savings to maintain the target cost of the new product. In the Komatsu case, the conflict between quality and cost was resolved by changing the way the ripper mounting bracket was attached to the bulldozer. (The ripper mounting bracket enabled the ripper to be attached to the main frame.) The new approach allowed the mounting bracket to be welded instead of bolted to the main frame. Welding was cheaper than bolting, and the savings equaled the additional cost of adopting the alternative design of the engine, transmission, and torque converter.

This example highlights an important observation. Target costing does not set out to identify the lowest cost at which a product can be manufactured but instead specifies a target cost that has to be achieved. Under target costing, engineers are not motivated to find cost savings simply for the sake of reducing costs. It is only when a design change has caused total costs to exceed target that the design engineers are motivated to search for additional cost savings.

Refusing to Launch Unprofitable Products

The second way the cardinal rule is enforced is by refusing to launch products that will not generate adequate profits. That is, products that cannot be manufactured at their target costs are not allowed to enter manufacturing. This application of the cardinal rule is what gives target costing its ultimate authority. At Sony the product planners do not have absolute freedom in relaxing a product's target cost. As a matter of policy, Sony will not sell products at a loss and, under most conditions, will not sell them below the minimum profit margin established by the appropriate business group manager.

Similar rules are observed at other companies. For example, at Citizen Watch the final step in the product review is to estimate the profitability of the new watch by subtracting the expected costs from the selling price and multiplying the result by the anticipated volume. If the watch is profitable it is introduced, and orders are accepted from Citizen Trading Company and other customers. If the watch is unprofitable, then the selling price, production cost, and design are reviewed. If there is no way for the product to be made profitably, it is never introduced. The primary motivation for these rules is to maintain the profitability of the firm.

Managing the Transition to Manufacturing

Once the target cost has been established and the product design completed, the product is released to manufacturing. As part of the transition from product design to manufacturing, the target cost is compared to the standard cost of the product. This comparison is necessary because there are differences between target and standard costs. A target cost is an expected cost set during the design stage of a product's life. A standard cost, in contrast, is the expected (or prior actual) cost of the product during its manufacturing stage. They are not the same thing. The target cost is established before all the details of production are known, and so it does not reflect current conditions such as the going rate for labor and materials or the specific plant in which the vehicle will be manufactured. In contrast, the product's standard cost is adjusted for all these factors.

The product's standard cost is expected to be equal to or lower than its target cost. If it exceeds the target cost, the firm takes action to reduce manufacturing costs until it reaches the target cost. At Nissan, as the vehicle enters production, accounting monitors all component and assembly costs. If these costs are not in line with the final target costs, accounting notifies cost design and engineering, and that department then performs additional value engineering to reduce costs to the target levels.

In many settings, once the manufacturing cost is equal to or lower than the target cost, product-specific cost-reduction activities cease (though general kaizen activities continue). The logic behind this cessation is that the costs of disrupting the production process exceed any potential benefits. Nissan, for example, does not undertake any cost-reduction efforts during the production stage unless the production cost exceeds the target cost. Management has determined that the incremental savings from such efforts are more than offset by the disturbance to the production process. When inflation or other factors cause costs to rise, the suppliers are pressured to find ways to keep component costs at their final target levels. Similarly, pressure is exerted on the assembly plants to achieve the assembly target costs.

Allowing for Extenuating Circumstances

Sometimes firms appear to violate the cardinal rule by releasing a product whose selling price is insufficient to cover both manufacturing costs and target profits. These exceptions are not necessarily violations of the cardinal rule but rather instances where a narrow, individual-product perspective is deemed inadequate. Typically, these products have future revenues associated with them that are not captured in their individual selling prices. If these additional revenues are included in the target cost derivation, the products do meet the requirements of the cardinal rule. At Sony, the only exceptions to the cardinal rule are strategic products expected to pay off in the long run, because management views them as investments necessary to create or expand markets.

Common examples of products that have associated future revenues are flagship products, products that use the next generation of technology, and products that protect market share. Flagship products create market awareness of the firm's name and presumably lead to increased sales of other products. Citizen considers the perpetual calendar watch a flagship product. Given its role, it is not necessary for the flagship product itself to be profitable, but rather it helps increase the overall profitability of the firm.

Products that use the next generation of technology are protected because they potentially hold the key to the future of the firm. For example, Olympus manufactures and sells electronic still cameras. These products currently sell at a loss. They are not abandoned, however, because it may well be that in 10 years electronic as opposed to chemical film cameras will dominate the photography market. If Olympus has not developed proprietary technologies for electronic still cameras, it will be at risk of having to leave the camera market. The potential future revenues and profits from electronic still cameras thus offset the need to be profitable today.

Finally, certain products play a strategic role in the product line. The least expensive cameras Olympus sells are entry-level ones. They are the first cameras a customer buys or is given as a present. Once someone has bought an Olympus camera, he or she typically will buy a new camera every few years. As long as Olympus can keep these customers satisfied, the future cameras they buy will in all likelihood be made by Olympus. The starter cameras have an associated future revenue from the replacement cameras, so even if a starter camera cannot be produced at its target cost, it will be launched if Olympus believes that failure to do so would cause a large customer loss in the long run.

When it is beneficial to the firm, even products that do violate the cardinal rule may be launched. Typically, the rationale supporting the release of these products centers on lost sales. The highly competitive markets in which these firms compete often do not allow product release to be delayed for more than a few weeks. Sony believes that it is imperative to release products on a timely basis and doesn't allow product redesign to extend the launch date. The Walkman market is so competitive that failure to release a new model timely typically will result in considerable lost sales. Because the physical production facilities still exist, the firm sees no benefit in missing a launch date, and it will launch a product to meet deadlines even if its profitability is below the minimum level.

When such a product is launched, it undergoes two analyses: a thorough review of the design process to identify why the target cost was not achieved and an intense cost-reduction effort immediately after its launch so that the violation of the cardinal rule is as short lived as possible. The purpose of these two analyses is to maintain the discipline of target costing even though a temporary violation has occurred.

ACHIEVING THE TARGET COST

Once planners have identified the target cost-reduction objective, the next task becomes finding ways to achieve it. Several engineering techniques can help product designers find ways to reduce the costs of products. They include value engineering (VE), design for manufacture and assembly (DFMA), and quality function deployment (QFD).

Value engineering has the primary objective of maximizing customer value—it tries to increase functionality and quality while at the same time reducing cost. In contrast, DFMA focuses on reducing costs by making products easier to assemble or manufacture, while holding functionality at specified levels. Finally, QFD, which is a visual, decision-making procedure for multiskilled project teams, provides a structured approach to ensure that customer requirements are not compromised during the design process. Of these three techniques, value engineering is the most important because it deals with all three characteristics of the survival triplet simultaneously. It will be discussed in greater depth in Chapter 7.

SUMMARY

Product-level target costing consists of three major steps. The first is setting the target cost of future products. The second consists of disciplining the target costing process. Achieving the product-level target cost is the final step. Product-level target costs are required because the derivation of allowable costs does not take into account the manufacturing realities of the firm. When the allowable cost is considered unachievable, it is increased to give the target cost. This target cost is designed to be achievable but only with considerable effort to reduce costs.

Comparing the target cost to the current cost of the new model identifies the target cost-reduction objective for the new model. The difference between the allowable cost and the product-level target cost represents the strategic cost-reduction challenge. This challenge identifies the inability of the firm to achieve strategic parity with its competitors.

While the current cost assumes no cost-reduction improvements, the as-if cost identifies the cost of the product if it were manufactured today, assuming incorporation of all known improvements. The drifting cost is the estimate of what the product cost would be if the product were manufactured at any point during the product design process, given confirmed cost savings. The drifting cost has to shift from the current cost to the target cost by the time the product design process reaches completion. Monitoring the relationship between the drifting cost and the target cost is an important disciplining mechanism. For firms that are driving toward perfection, the unavoidable waste-free cost and the perfect waste-free cost are better benchmarks against which to measure their progress than looking at their competitors' performance.

The cardinal rule of target costing plays an important role in maintaining the discipline of target costing. Great care is taken to ensure that the sum of the component-level target costs does not exceed the target cost of the product. Often, an increase in the cost of one component causes the engineers to explore ways to reduce the costs of other components by an equivalent amount. In addition, to help ensure enforcement of the cardinal rule, most firms have a policy against launching unprofitable products.

When the product design phase is over, the product moves to manufacturing. As part of this transition phase, the target cost is compared to the standard cost of production. If the standard cost is higher, then usually the firm takes steps to reduce manufacturing costs to the target level. Often if the standard cost is at or below the target cost, the design of the product is frozen for the rest of its life, and no further actions, other than general kaizen, are taken to reduce the cost of the new product.

As with any rule, the cardinal rule occasionally is broken. It is violated when a broader analysis indicates that doing so will be beneficial for the firm. Target costing, by its nature, takes a single-product orientation. Sometimes this view is too restrictive because the product under review causes additional revenues to be generated beyond those generated by the product itself. Such products include flagship products that create high visibility for the firm, products that use the next generation of technology, or products that fill a critical gap in the product line. For such products, the target cost is often relaxed to allow for the "hidden" revenues.

At the heart of the process of achieving the product-level target cost is value engineering, a structured approach to product design. VE focuses on the functionality of products and components and ways to achieve desired levels of functionality at an acceptable cost. It is the subject of the next chapter.

CHAPTER 7

VALUE ENGINEERING

INTRODUCTION

Value engineering is a systematic, interdisciplinary examination of factors affecting the cost of a product so as to devise means of achieving the specified purpose at the required standard of quality and reliability at the target cost. In Japanese firms, the work force is usually organized into self-guided teams or groups, and it is these teams that actually achieve the firms' cost-reduction objectives. Consequently, the way in which these teams are motivated helps determine the success of the firms' cost-reduction programs.

Most value engineering (VE) occurs in the product-level and component-level sections of the target costing process, though a small part occurs during the price-driven costing section. The Isuzu case demonstrates the richness of practice of a mature, fully developed VE program. At Isuzu, this program consists of numerous subprograms, each designed to create a functionality-cost analysis either at different stages of the product development process or for different cost elements.

The initiation of VE marks the transition from setting target costs to achieving the required cost reductions. The key to successful cost management lies in the existence of a committed and motivated work force. Without the right organizational context simply launching cost-reduction programs will not work. Thus, the organizational context plays a critical part in the success of VE programs.

THE ORGANIZATIONAL CONTEXT

At most firms, the team leaders (typically high-school graduates) function like managers and act with a high degree of autonomy. Team members are often drawn from different parts of the firm. Since successful cost reduction must balance all three characteristics of the survival triplet, multifunctionality is critical. At Topcon, the MAST (management activity by small team) program consists of cross-functional, self-directed teams formed to achieve a specific objective. A typical objective for an accounting MAST team is to find ways to improve the accuracy of the inventory records.

MAST teams operate at the division manager level. The team leader forms a cross-functional team typically from manufacturing, production, accounting, marketing, and, in particular, the TQC and value analysis departments. Usually each manager is a leader of one team and a member of several others. This interlocking membership allows the teams to be aware of the actions of other teams and so avoid duplication of effort. Similarly, at Nissan the allowable cost is set by teams derived from almost every functional area of the firm, including product design, engineering, purchasing, production engineering, manufacturing, and parts supply.

Everyone is involved in cost management because every team has a cost-reduction target. In most firms, the procedure for setting team cost-reduction targets is part of a hierarchical target-setting process. In companies that use a top-down approach, upper management sets the targets. The process begins with corporate-wide cost-reduction targets set during the annual planning process and distributed among the divisions. At this stage, negotiations over the targets are between the corporate planning department and the divisional managers. In the next stage, the divisional cost-reduction targets are distributed among the production facilities in the divisions and then to the teams.

The top-down approach does not always create commitment, however. At Olympus, they discovered that over the years, groups' initial cost-reduction targets had become biased downward to create slack that could be used to help the groups achieve their negotiated targets. When divisional management became aware of this practice, it would take it into consideration as much as possible when establishing new targets.

In companies using a bottom-up approach, team leaders are responsible for setting their own targets. At Olympus, for example, cost-reduction targets are identified for each product produced by the group. The group leaders recommend their cost-reduction targets, and divisional management reviews these recommendations. If the overall savings are insufficient, the targets are renegotiated until divisional management finds the savings acceptable.

While team leaders have considerable autonomy, it is conditional on their success. If a team fails to achieve its target, management takes a series of steps to correct the problem. At Olympus, the team leader and foreman in each group meet daily to discuss their progress. Group and team leaders hold weekly meetings to report on progress. If a group doesn't meet its weekly objectives, the group leader is expected to explain why the group failed and what corrective actions will be taken. Occasionally, if a group consistently fails to meet its objectives, management sends in engineering—a serious blow to the group's reputation.

The role these small teams play in achieving the cost-management objectives is critical. Without them, the grassroots commitment to cost reduction would not occur. It is the commitment of the groups to the cost-reduction objective that creates the fundamental pressure for cost reduction.

PLANNING THE VE PROCESS

The application of value engineering begins with the conceptualization of the product and continues through the design process until the product is released to manufacturing. Even then the process continues, but under the name value analysis (VA). The difference between VA and VE is not in the approach taken or tools used but in the point at which they occur in the life cycle of the product. VE is used during the product design and development stages, while VA is used for the manufacturing stage and for purchased parts. For this reason target costing and value engineering can be viewed as concurrent activities as can kaizen costing and value analysis.[1]

It would be wrong to view VE as just another cost-reduction program. VE primarily focuses on product functions and only secondarily on cost. The motivating force behind VE is to ensure the product achieves its basic function in a way that satisfies the customer at an acceptable cost. Consequently, VE programs are the domain of the product engineer, not the accountant.

We chose Isuzu to demonstrate the practice of VE because it is widely recognized as both a pioneer in VE methods and one of its most sophisticated practitioners. To help achieve its cost-reduction objectives and ensure that a product meets its target cost, Isuzu relies heavily on a cost deployment flowchart and a cost strategy map. The firm uses the flowchart to ensure that cost-reduction activities are applied to a product as early as possible in its development. The chart came about early in their VE experience because Isuzu management realized that no significant analysis had been performed during the product-planning stage to ensure that its target cost could be achieved. When it finally became clear that the product could not be manufactured for its target cost, it was too late in the development process to reduce costs effectively and economically.

The cost deployment flowchart identifies five development stages in a vehicle's design: concept proposal stage, planning stage, development and product-preparation stage, development and production/sales-preparation stage, and production/sales-preparation stage. The flowchart also indicates the type of cost-reduction activities required, the steps involved in each activity, the division responsible for the activity, and the development stages at which each activity should occur. For instance, during the concept-proposal stage, management determines a new vehicle's target cost, performs teardown analysis, and holds a design review.

---

[1] Kaizen costing and value analysis will be the subjects of the third volume in this series.

In addition, to ensure consideration of the appropriate cost-reduction techniques at the right time throughout the development process, Isuzu developed a more detailed approach, the cost strategy map. This map is needed because the appropriate technique depends not only on the stage of the development process but also on the part in question. Some cost-reduction techniques such as first- and second-look VE are suitable only for making drastic changes to the design of a part; they are not suitable when an existing design is being slightly modified.

The cost strategy map is prepared after the cost deployment flowchart. It lists the applicable cost-reduction techniques for each development stage in more detail than the flowchart. To illustrate, in the planning stage, the cost deployment flowchart identified two major blocks of cost reduction: first-look VE and a mixture of first- and second-look VE and teardown. The corresponding cost strategy map identified 12 techniques; 8 were teardown methods, and the other 4 were types of first- and second-look VE.

The cost deployment flowchart and the cost strategy map were used together to ensure that the appropriate cost-reduction techniques were applied as soon as possible in the development of new products. To maximize cost reduction, Isuzu had created a comprehensive cost-reduction program consisting of a large number of different cost-reduction techniques. Some of these techniques were developed by Isuzu. Others were adapted from systems developed at other firms.

VE Techniques

The VE techniques illustrated in the Isuzu case can be broken into three major categories: three Nth-look approaches, eight teardown approaches, and six other VE approaches.

The Three Nth-Look VE Approaches

The three Nth-look VE approaches are designed to be applied at different stages of the product design process. Zero-look VE is the application of VE principles at the concept proposal stage, the earliest stage in the design process. Its objective is to introduce new forms of functionality that did not previously exist. First-look VE focuses on the major elements of the product design and is defined as developing new products from concepts. The objective is to enhance functionality of the product by improving the capability of the existing functions. Second-look VE is applied during the last half of the planning stage and the first half of the development and product preparation stage. The objective of second-look VE, unlike that of the zero- and first-look, is to improve the value and functionality of existing components, not create new ones. Consequently, the scale of changes is much smaller than for zero- and first-look VE.

Zero-Look VE

Zero-look VE is a recent introduction. It represents the logical extension of VE into the earliest stage of the design process, the concept proposal stage, when the basic concept of the product is developed and its preliminary quality, cost, and investment targets are established. Unlike first-look VE, which acts to enhance the functionality of a product by improving the capability of existing functions, the role of zero-look VE is to introduce some forms of functionality that did not previously exist. While the underlying concept of designing revolutionary products has always been part of VE, codifying the VE process as zero-look at this stage makes VE an integral part of product design and increases the likelihood of its occurring.

Zero-look VE was applied in the development of Isuzu's NAVI-5 transmission system, which combines the higher fuel efficiency and performance of a manual transmission with the convenience of an automatic transmission. In simplified terms, NAVI-5 is a computer-controlled manual transmission capable of changing gears automatically. It was zero-look VE that first identified the basic concept behind the new system.

First-Look VE

First-look VE, which is more traditional and is used by numerous Japanese firms, focuses on the major elements of the product design and is defined as developing new products from concepts. It is applied during the last half of the concept-proposal stage and throughout the planning stage. In the planning stage at Isuzu, the planners identify the key components or major functions, determine the commodity value (i.e., the product's type, quality, size, price, and function), submit a design plan, distribute target costs to the vehicle's major functions such as its engine, transmission, and air conditioner, and set the degree of component commonality. They use first-look VE at this stage to increase the value of the product by increasing its functionality or reducing its cost.

For example, Isuzu's engineers applied first-look VE to the development of the Gemini heater. They determined that reducing the time it took for the automobile interior to warm up would be a benefit that users would welcome. Consequently, they initiated a project to find ways to heat the car interior before the engine warmed up. The ultimate solution was to install a ceramic heater that functioned only when the engine was below a specified temperature. This heater was used to warm the air flow directed toward the occupants' feet. When the water in the engine reached the specified temperature, the ceramic heater switched off and the traditional heater took over.

Second-Look VE

Japanese firms also widely use second-look VE. They apply it during the last half of the planning stage and the first half of the development and product-preparation stage. In the development and product-preparation stage, the components of the main functions are identified and handmade prototypes are assembled. At Isuzu, second-look VE was applied on the ELF, a light-duty truck. Experience with earlier models had shown that the gear lever, which was positioned between the two front seats, sometimes got in the way of the occupants. Repositioning the gear lever so that it was out of the way would improve the vehicle's functionality and hence its value. The solution was to develop a gear lever that could fold down while the vehicle was stationary but that would not collapse while the vehicle was in motion.

The Eight Teardown Approaches

Isuzu's approach to teardown indicates the firm's large range of cost-reduction activities. It uses teardown methods to analyze competitive products in terms of the materials they contain, the parts they use, the way they function, the way they are manufactured, the way they are coated, and the types of coating used. Isuzu defines the teardown method as *"a comparative VE method through visual observation of disassembled equipment, parts, and data arranged in a manner convenient for such observation."*

Teardown methods were introduced in 1972, the year after General Motors (GM) purchased a 37% share of Isuzu. The GM method (or static teardown method, as it was known at Isuzu) was modified subsequently over a three-year period to fit Isuzu's needs and then was allowed to proliferate throughout the company. The primary difference between the original GM approach and the Isuzu method was the scope of application of teardown principles.

Isuzu uses teardown methods in all stages of product development. The firm has eight different teardown methods: dynamic, cost, material, static, process, matrix, unit-kilogram, and group estimate. The first three methods are designed to reduce a vehicle's direct manufacturing cost. The next three are intended to reduce the investment required to produce vehicles through increased productivity. The last two techniques are an integration of teardown and VE techniques.

Other VE Techniques

In addition to zero-, first-, and second-look VE and the eight teardown approaches, Isuzu uses four other cost-reduction techniques: the checklist method, the one-day cost-reduction meeting, mini-VE, and the VE reliability program.

The checklist method is used to identify a product's cost factors and to suggest ways to reduce costs. The checklist consists of a number of questions designed to guide the firm's cost-reduction activities by discovering cost-reduction opportunities. Checklists help ensure exploration of all possible avenues for cost reduction.

The firm holds one-day cost-reduction meetings to improve the efficiency of the entire cost-reduction process, including VE and teardown methods. Participants from engineering, production, cost, and sales are expected to come up with ideas for new cost-reduction possibilities. The meetings are a way to overcome limitations in the approval process used for most cost-reduction proposals. The approval process entails circulating written proposals to all involved parties, who indicate acceptance by signing off on them. Unfortunately, this approach severely reduces the exchange of information and modification of ideas. At the one-day meetings, presentation of the results of various teardown programs helps initiate discussions.

Mini-VE is a simplified approach to second-look VE. It is applied to specific areas of a part or to very small, inexpensive parts. Isuzu applies this process to the design of mirrors, doors, and door locks. For example, as a result of mini-VE, product mirrors may be redesigned to be more ergonomic. Mini-VE is applied during the development and product preparation stages, the development and production-sales preparation stage, and the production-sales preparation stage.

The VE reliability program is designed to ensure that the most appropriate form of VE is applied to each problem. It is essentially a "quality of VE" program. For example, if a completely new product design is required, applying second-look VE is not appropriate. Like mini-VE, the program is applied during the development and product-preparation stages, the development and production/sales -preparation stage, and the production/sales-preparation stage.

SUMMARY

The primary purpose of VE is to increase the value of a firm's products, where value is defined as the functionality of a product divided by its cost. VE is a multifunctional discipline that analyzes products in terms of their basic and secondary functions. A product's basic function is the principal reason for its existence. Secondary functions are outcomes of the way the designers choose to achieve the basic function.

When VE is integrated with a target costing system (which holds the cost of each product constant), its objective is to increase the functionality of products while maintaining their target costs. A deep understanding of perceived value is critical if the firm's VE programs are to increase the value of products and keep these products within their survival zones.

The way in which firms increase the value of their products using VE can be quite complex, and Japanese firms have developed numerous variations of VE techniques. Isuzu, for instance, uses three stages of VE (zero-look, first-look, and second-look) as well as eight different teardown approaches and four other cost-reduction techniques to improve the value of its models.

The three VE approaches focus on different phases of the design process. Zero-look focuses on the earliest phase of product design. Its objective is to introduce new forms of functionality. First-look VE focuses on the conceptual phase of product design and works to enhance the functionality of new products. Second-look VE focuses on the last phases of product planning and attempts to find ways to improve the functionality of existing components. Although each of the eight teardown methods focuses on a different objective, they all rely to some extent on the knowledge gained when a product (either Isuzu's or a competitor's) is taken apart so its constituent parts and the way they are assembled can be studied.

Other VE programs to reduce a new product's cost or functionality are the checklist method, one-day cost-reduction meetings, mini-VE, and the VE reliability program. Finally, to help guide the entire VE process, Isuzu developed the cost deployment flowchart and cost strategy map. These two documents provide guidelines as to which VE techniques to use at each of the different phases of product design.

The Isuzu case does not provide an exhaustive list of VE techniques, but it does illustrate the richness of practice and the importance of VE to the target costing process. Without a commitment to achieve target costs and an aggressive VE program, firms will find it difficult to survive in a confrontational setting. Inherent to the VE process is a rich interaction with the firm's suppliers. This interaction is part of the component-level target costing process, the subject of the next chapter.

CHAPTER 8

COMPONENT-LEVEL TARGET COSTING

INTRODUCTION

The completion of the product-level target costing part of the target costing process signals the beginning of the third and final part, *component-level target costing* (see Figure 8-1). This part of the process decomposes the product-level target cost to the component level. The component-level target costs identify how much the firm is willing to pay for the components it purchases.

Thus, component-level target costing sets the selling prices of the components manufactured by the firm's suppliers and focuses supplier creativity on finding ways to design components at low cost. It uses *interorganizational costing* to achieve this objective by opening new communication channels among suppliers, customers, and product designers.

The third section of the target costing process can be broken into four steps (see Figure 8-2). The first step is to set the target costs of major functions, and the second step is to set the component-level

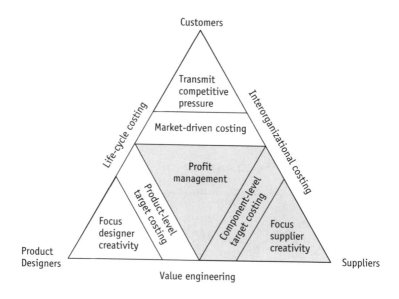

target costs. The next step is to select the suppliers for the components. The final step is to reward suppliers for their creativity.

SETTING THE TARGET COSTS
OF MAJOR FUNCTIONS

Once the firm has established the target cost of a product, it develops target costs for the product's components. This process enables the firm to achieve the second objective of target costing: transmitting the competitive cost pressure the firm faces to its suppliers. This objective is critical in lean enterprises, which are horizontally rather than vertically integrated. Such firms purchase a significant portion of the parts and materials required to manufacture their products from external instead of internal suppliers.

For example, in 1993, Komatsu manufactured about 30% of the material content of its products, designed and subcontracted another 50%, and purchased the remaining 20% from outside suppliers. At Toyota, third-party suppliers are responsible for approxi-

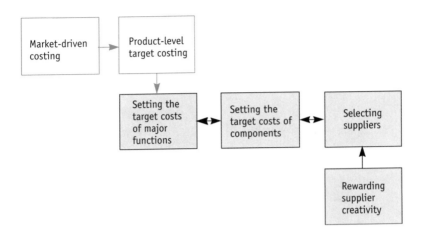

FIGURE 8-2. COMPONENT-LEVEL TARGET COSTING

mately 70% of the parts and materials required to produce the firm's automobiles. This high level of dependency on externally supplied items makes supplier relations extremely critical to the firm's success. In particular, the cost and quality of third-party-supplied parts is considered critical.

Thus, the horizontal integration that allows lean enterprises to be highly flexible and responsive creates a heavy reliance on suppliers. Target costing provides a powerful mechanism to discipline suppliers by allowing the firm to set the selling prices of the components they supply. When products are complex, the process of setting target costs for externally acquired components is often carried out by first establishing the target costs of the major functions and then, in a separate step, of the components they contain (see Figure 8-3). Major functions are the subassemblies that provide the functionality that enables the product to achieve its purpose.

At Isuzu, for example, the designers identify approximately 30 major functions per vehicle, including the engine, transmission, cooling system, air conditioning system, and audio system. Group components are the major subassemblies purchased from the firm's suppliers and subcontractors. There are only about 100 such components, yet they amount to as much as 70% to 80% of the manufacturing cost. Group components include the carburetor and starter.

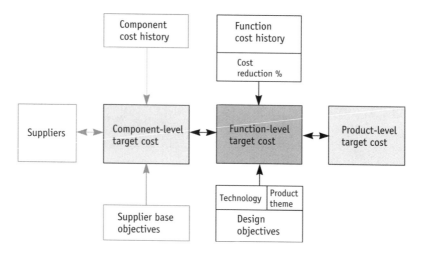

FIGURE 8-3. SETTING THE TARGET COSTS OF MAJOR FUNCTIONS

The Role of Design Teams and the Chief Engineer

Identifying major functions allows the design process to be broken into multiple, somewhat independent tasks. Typically, the design of each major function is the responsibility of a dedicated team. Design teams usually include representatives from a number of disciplines such as product design, engineering, purchasing, production engineering, manufacturing, and parts supply.

Most of the firms, including Toyota, organize product design around a matrix structure with each design team reporting to the chief engineer responsible for the product and the head of the appropriate design division. The purpose of the matrix structure is to balance the unique requirements of each product with the desire to maintain common design philosophies across products (see Figure 8-4).

The overall responsibility for coordinating the design of a new product typically rests with the chief engineer or product manager, who selects the distinctive theme of the new product and sets its functionality. While playing a critical role in the development of a new product, usually the chief engineer is supported by a relatively small team of design engineers. At Toyota, with the help of engineers in the design, test-production, and technical divisions, a chief engineer drafts the development plan for the new model and then leads the development project. Well over 100 engineers from the various divisions work with a chief engineer on a typical project, but since they belong to different divisions, not all team members are under his supervision. Probably only about a dozen people report directly to him. In this sense, the chief engineer is more a project leader than a supervisor of product development.

This matrix approach has several advantages. First, the chief engineers are responsible for coordinating the design process at the design divisions. The design divisions are relatively autonomous, and chief engineers are expected to develop a concept for the new vehicle that spans multiple design divisions. Keeping the design divisions autonomous is considered important, as it allows sharing of expertise across all design projects within each division. The firm feels that the tensions created by this matrix approach are beneficial to the creative design process and worth any conflict they create.

Most of the firms set different cost-reduction objectives for each major function (see Figure 8-5). At Toyota the purpose of cost planning is to determine the amount by which costs can be reduced through better design of the new model. The cost-planning goal is distributed to the divisions in charge of design for the model, with

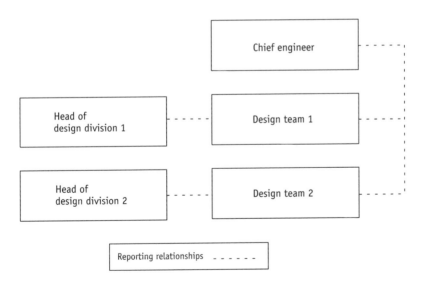

FIGURE 8-4. THE TOYOTA MATRIX STRUCTURE

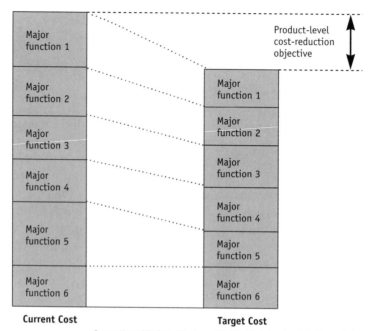

Source: Harvard Business School case 194-040. Adapted and reprinted by permission.

FIGURE 8-5. NISSAN MOTOR COMPANY, LTD.: DECOMPOSING
THE TARGET COST TO THE MAJOR FUNCTION LEVEL

each division being given a particular cost-reduction target. For example, the divisions in charge of design of the engine, body, chassis, drive train, electronics, and interior all receive different cost-reduction targets.

Toyota management believes that it is impossible to attain a target cost simply by decreeing a uniform reduction of costs for all divisions. Consequently, the chief engineer distributes a portion of the goal to each design division. Discussion continues with each division until both the division and the chief engineer are satisfied with the amount distributed, which is based on precedent and experience. The divisions are held responsible for attaining their cost-reduction goals.

The Target Cost-Setting Process

The chief engineer is responsible for setting the target cost of each major function, usually through an extended negotiation process with the design teams. The target costs typically are based on historical cost-reduction rates. If the cost of a major function historically has been decreasing by 5% a year, then this is the rate that usually will be used.

Not all firms rely solely on historical cost-reduction rates. Some use market analyses to help set the target costs of new products. These market-based approaches are particularly applicable when new forms of product functionality are being introduced. For example, Isuzu uses monetary values or ratios to help set the target costs of major functions and asks customers to estimate how much they are willing to pay for a given function. These market-based estimates, tempered by other factors such as technical, safety, and legal considerations, often lead to adjustments to the prorated target costs. For example, if the prorated target cost for a component is too low to allow a safe version to be produced, the component's target cost is increased, and the target cost of other components is decreased to compensate.

Toyota adopts a different approach than the other firms. Rather than adding together all the costs for a new model, Toyota's approach to cost planning is to sum the differences in cost between new and current models. At Toyota the estimated cost of a new model is described as the cost of the current model plus the cost of any design change. Thus, for every increment in the functionality of a new model there is an estimated incremental price and cost. This approach allows the firm to measure the incremental profitability of each new function it builds into a new model.

This approach has several advantages. Cost planning can begin even before blueprints for the first test model are drawn. Estimating the total difference instead of the total cost tends to be less troublesome and more accurate, and this method helps the related divisions understand cost fluctuations. Toyota considers the differences approach more accurate because the typical new model is based heavily on existing designs. Trying to estimate the cost of a new vehicle from scratch would, in management's opinion, introduce more errors than using existing data and modifying them accordingly. The approach is also more helpful to the design divisions because it highlights the areas of the new model that are different from existing designs. It is these new designs that require the most work in the design divisions.

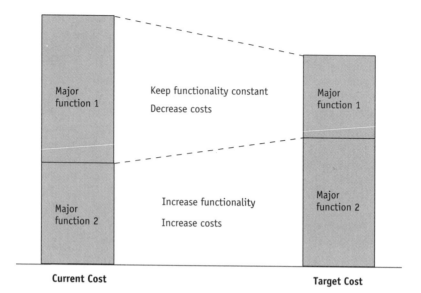

FIGURE 8-6. DISTRIBUTING THE TARGET COST
ACROSS MAJOR FUNCTIONS

The chief engineer will modify the target costs derived either from historical rates or market analysis for three major reasons. First, if the sum of all the historical rates doesn't give the desired product-level cost-reduction objective, the chief engineer will negotiate with the head of the design teams of the major functions for higher rates of cost reduction. These negotiations continue until the sum of the target costs of the major components equals the target cost of the product.

Second, if the relative importance of the major function changes from one generation to the next, the chief engineer will modify the target costs accordingly. For example, if he wants the new vehicle to be quieter and sportier, he might increase the target cost of those major functions to make it easier for the design team to achieve both their functionality objective and target cost (see Figure 8-6). At Toyota, the chief engineer is expected to make his own decisions about where cost reduction is to occur. One of the objectives of the target costing system is to focus the attention of the design engineers in the design divisions in the right place. The chief engineer typically has objectives for the new vehicle that affect where costs can be reduced. To achieve these objectives, he will decrease the cost reductions expected from the design divisions responsible for those aspects of the product and increase the expected reductions from other divisions because under the cardinal rule of target costing, these cost increases have to be offset elsewhere in the design (see Figure 8-7). At Nissan, although the target cost (of a major component) is usually lower than the current cost, sometimes the allowable cost is higher because the new product specifications demand higher performance and functionality than existing designs. In total, the sum of the cost reductions for each major component is meant to equal the required level of cost reduction to achieve the model's allowable cost.

Third, when the technology on which the major function relies changes, the historical cost-reduction rate of the old technology ceases to be meaningful. The historical rate for the new technology should be used instead, if it is available. When entirely new technologies are used, the cost estimation problem becomes more difficult because no historical data on cost-reduction trends have been developed.

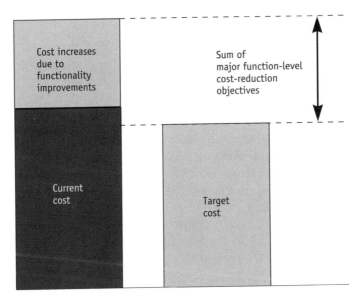

FIGURE 8-7. ACHIEVING THE TARGET COST

At some Japanese firms, a safety factor, often called "the reserve for the production manager" is built into the development of component-level target costs. The purpose of this reserve is to allow for any (minor) cost overruns due to design-related problems that may occur during the production process. Such a reserve is advantageous because experience has shown that overruns are common in practice and cannot easily be avoided. If the sum of the component-level target costs is set to be equal to the product-level target cost, then such overruns will cause the product to violate the cardinal rule and exceed its target cost. By establishing an adequate reserve, the number of minor cardinal rule violations can be significantly reduced without reducing the overall discipline of target costing.

The chief engineer is responsible for creating a reserve that is sufficient to offset the anticipated overruns but that does not introduce slack into the production process. The magnitude of the reserve is based on previous experience from similar product development projects. The reserve is usually quite small, ranging from 5-10% of the product-level target cost. It is determined by setting the component-level target costs and assembly and indirect manufacturing target costs so that they sum to an amount that is below the product-level target cost (see Figure 8-8). That is:

Product-level target cost – sum of component-level target costs
– assembly target costs
– indirect manufacturing target costs
= reserve for production manager

The creation of the reserve introduces another step into the component-level target costing process. This step occurs before the component-level target costs are set, when the chief engineer must decide how large a reserve to create (see Figure 8-9). While the reserve is spread across the major functions and components that the product contains, no attempt is made to identify the reserve below the product level. The purpose of the reserve is to allow for cost overruns at the product level, not at the major function or component level. If the reserve was used to enable cost overruns at the major function and component level to be absorbed, it would create slack in the target costing process. It would reduce the effec-

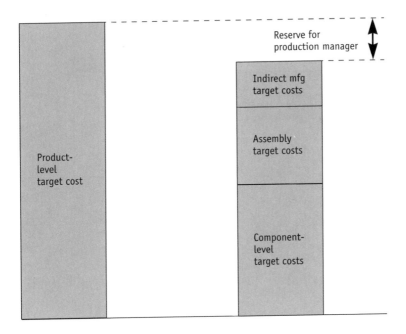

FIGURE 8-8. THE RESERVE FOR THE PRODUCTION MANAGER

tiveness of the cardinal rule and hence dilute the discipline associated with target costing.

When the product enters manufacturing and the anticipated cost overruns occur, they eat into the reserve, as opposed to violating the target cost. As long as the overruns do not exceed the reserves, the product will achieve (or be slightly under) its target cost. If all the reserve is not required, the remainder can be used to add additional functionality and bring the cost of the product up to the product-level target cost, or it can be used to increase the advertising spent on the product. It usually is not used to increase the profitability of the product. If the overruns exceed the reserve, then the product violates its target cost and is either abandoned or subject to immediate, product-specific kaizen costing to bring its costs back in line.

Once the target costs of the major functions have been established, they are decomposed to the group component and parts level as appropriate. The objective is to set a purchase price for every externally acquired group component or part.

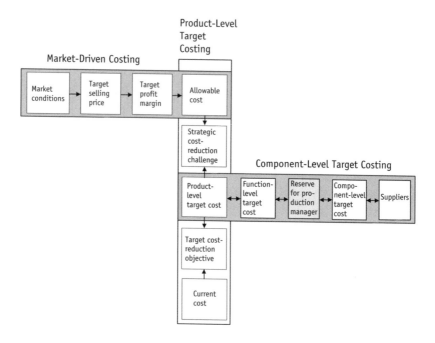

FIGURE 8-9. THE TARGET COSTING PROCESS REVISED
FOR THE RESERVE FOR THE PRODUCTION MANAGER

SETTING THE TARGET COSTS
OF COMPONENTS

Target costs for components can be set only when the product design has reached the stage at which specific components can be identified. For example, at Nissan value engineering is used after the engineering drawings for trial production are complete, to determine target costs for each of the components in every major function of the automobile. These estimates are derived by identifying a cost-reduction objective for each component.

The process of setting component-level target costs is similar in many respects to that used to set the target costs of the major functions. The component-level target costing process consists of three major blocks (see Figure 8-10). The first uses the component cost history as the starting point for estimating the new component-level target costs. The second applies the firm's supplier-base

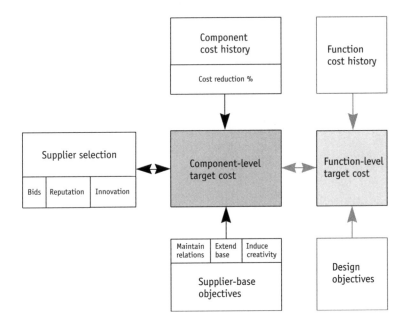

FIGURE 8-10.   SETTING THE TARGET COSTS OF COMPONENTS

objectives to the selection of suppliers in general. The third deals with the selection of the supplier for a given component. Component-level target costs are set primarily by the design teams and occasionally by the chief engineer.

The Role of the Design Teams and the Chief Engineer

Typically, it is up to the major function design teams to decompose the target cost of the major function to the component level (see Figure 8-11). However, sometimes the chief engineer gets involved in the process to ensure that his objectives for the product are met. At Toyota, the cost-reduction targets are assigned first to the design divisions and then for certain major parts decomposed to the part level. For example, if one of the design changes is to increase the power of the engine by 10hp, the engine division might estimate that the improvement will increase the cost of the crankshaft

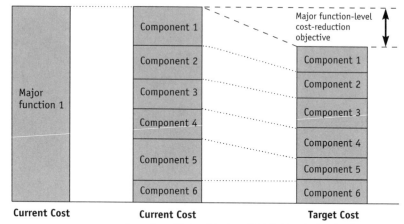

		Major function-level cost-reduction objective
	Component 1	
	Component 2	Component 1
Major function 1	Component 3	Component 2
	Component 4	Component 3
	Component 5	Component 4
		Component 5
	Component 6	Component 6
Current Cost	**Current Cost**	**Target Cost**

Source: Harvard Business School case 194-040. Adapted and reprinted by permission.

FIGURE 8-11. NISSAN MOTOR COMPANY, LTD.: DECOMPOSING
THE TARGET COSTS OF MAJOR FUNCTIONS
TO THE COMPONENT LEVEL

by ¥X. The chief engineer will use the precedents for upgrading engines by 10hp to estimate a more aggressive cost and then will ask the division to compromise on ¥Y. In contrast, another division might be asked to reduce costs because the new part will be smaller or lighter than the old one. A third division might be asked to maintain the same cost, despite a change in materials, because no change in performance is anticipated.

Body styling also has a major influence on cost. Some designs create complex part structures and require higher tolerances, thereby increasing the number of labor-hours required in production. For example, in the early 1990s, a trend emerged to make the bumper look as if it were an integral part of the body. The space between bumper and body has been reduced from several millimeters to less than a millimeter on recent models to achieve this objective. This change inevitably increased the required tolerances, thus increasing manufacturing costs. When a certain body style requires a cost increase, it is up to the chief engineer to decide how large an increase is acceptable.

Similarly, the chief engineer sometimes gets involved in setting the target cost for very expensive components. At Toyota each design division is responsible for attaining its respective cost-reduction goal. The specifics of parts, materials, and machining processes are left to their discretion. Exceptions are made for large, especially costly parts. The chief engineer will sometimes specify cost-reduction targets for specific parts to the related divisions. These specific-part targets are set at the same time as the divisional targets. Consider a part that is estimated to cost ¥3,000. If it is judged that a cost break on this particular part will contribute significantly to attaining the target goal for the entire model, the chief engineer may ask the related design division for a part-specific cost reduction of perhaps ¥500 (see Figure 8-12). Thus, the chief engineer plays a critical role in the design process by first setting the distinctive themes of the new product and then negotiating the target costs for the design teams. Once the target costs of the major functions have been accepted, the design teams can take over. They have two major challenges: to design the major function so that it can be manufactured or purchased at its target cost and to develop component-level target costs.

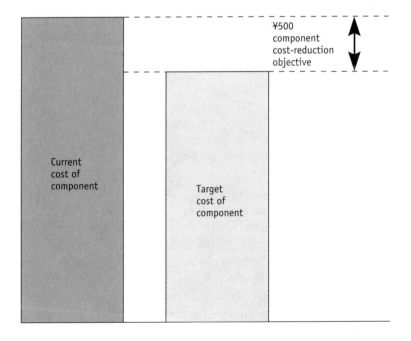

FIGURE 8-12.   SETTING THE TARGET COST OF A COMPONENT

Using Component Cost Histories

Several special techniques have been developed to set component-level target costs, including functional and productivity analysis. Komatsu uses these techniques to set the target costs of the components it acquires externally. The technique used depends on the way the product is sourced. Functional analysis, a procedure for identifying the target cost of a subassembly based on its functional characteristics, is used for parts designed and produced outside of Komatsu, such as cooling systems, hydraulic devices, and electrical subassemblies, because this procedure does not rely on detailed knowledge of the production process. Productivity analysis, a procedure for identifying the target cost of a subcomponent based on its manufacturing process, requires more in-depth knowledge of the production process and is therefore used for subassemblies designed by Komatsu (such as vehicle main frames, buckets, and gears) and manufactured either by Komatsu or by one of its subcontractors.

Functional Analysis

Functional analysis begins with an analysis of the primary and secondary functions of each subassembly and how they are achieved. Primary or basic functions are the principal reason for the existence of the product. For example, the primary function of the engine cooling system is to keep the engine within operating temperatures. The secondary functions are outcomes of the ways the designers choose to achieve the basic functions. The secondary functions of an engine cooling system include heating the passenger compartment.

The next step in the process is to identify how each subassembly, for example, the engine cooling system, achieves its primary function—cooling capacity. Often several determinants of primary functionality are identified and ranked in order of importance. These determinants typically relate to physical characteristics of the major components that make up the subassembly. (In the case of the cooling system these components include the radiator, fan, and electric motor.) The most important determinant of cooling capacity is the surface area of the radiator. The second most important determinant is the size of the fan, followed by its rotation speed, the volume of water in the system, and the ambient air temperature.

The third step in functional analysis is to use information about the subassembly components' functionality, for example, the radiator in the engine cooling system, and the component's physical characteristics, such as size, to identify the target physical characteristic (e.g., minimum size) of the component. The information for undertaking this analysis is contained in the firm's *functional tables.* The target physical characteristic is determined by plotting the capability of each major component in the subassembly (e.g., radiator) with respect to its primary functionality (for the radiator this is its cooling capacity) versus the most important determinant (for the radiator it is its surface area) for all similar components used by the firm.

From this plot, the average and minimum lines for existing components are constructed. The average line is determined using linear regression or other statistical estimation procedures, and the minimum line is drawn so that it passes through the most efficient of existing components. The minimum line and the required functionality for the new component are used to identify the target physical characteristics of the most important determinant of the functionality of the component (see Figure 8-13).

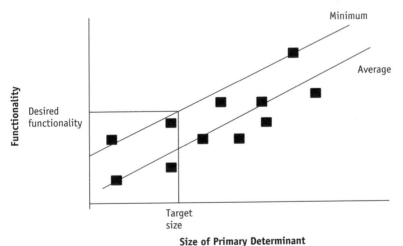

Source: Harvard Business School case 194-037. Adapted and reprinted by permission.

FIGURE 8-13. KOMATSU, LTD.: TARGET COSTING
            VIA FUNCTIONAL ANALYSIS

The target cost for each major component is determined by a similar process. This process compares the magnitude of the most important determinant (that is, physical characteristic) of the primary functionality of existing components (e.g., radiator surface area) to their cost. The appropriate cost and physical characteristic information is maintained in the firm's *cost tables*. The target cost of the components is determined by plotting the cost against their physical characteristics for all similar components used by the firm. The average line is determined using linear regression or other statistical estimation procedures, and the minimum line is drawn so that it passes through the most efficient of existing components. The minimum line and the target physical characteristic for the new component are used to identify its target cost (see Figure 8-14).

The same techniques are used to generate target costs for the other major components of the subassembly. For example, the target cost of the fan in the cooling system is determined by plotting the size of the fan against the cooling capacity of the engine cooling system and identifying the average and minimum lines for

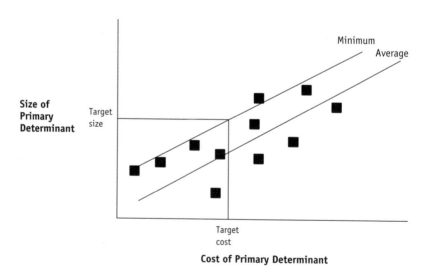

Source: Harvard Business School case 194-037. Adapted and reprinted by permission.

FIGURE 8-14.   KOMATSU, LTD.: TARGET COSTING
VIA FUNCTIONAL ANALYSIS—
IDENTIFYING THE TARGET COST

the relation between fan size and cooling capacity. The minimum fan-size-to-cooling-capacity line is used to identify the target fan size. A plot of the cost of fans versus fan size is used to determine the minimum cost/size line and hence the target cost of the fan.

This process is repeated for all major components of the subassembly. Once all component target costs have been identified, their sum can be compared to the target cost of the subassembly. As long as the cost of the components is less than or equal to the target cost of the subassembly, the components' target costs can be approved. Otherwise, additional analysis is required.

Productivity Analysis

Productivity analysis consists of decomposing the production process of the components of new products into detailed steps such as "drill a 10-mm diameter hole of 5-cm depth into stainless steel." Cost tables then are used to estimate the cost of each step in the production process. At Toyota cost estimators use cost tables to calculate unit prices for manufacturing. Cost tables are developed for five major production steps: machining, coating, body assembly, forging, and general assembly, and they detail the machine rates for each of the steps. These rates include labor, electricity, supplies, and depreciation costs.

The exact form of the cost table rate depends on the type of production step being analyzed; for example, for stamping. the cost table contains the cost per stroke while for machining it contains the cost per machine-hour. Cost tables are highly detailed, and each production line has its own cost table identifying, for example, the cost per stroke of each press. Rather than the basic costs used for budget management, a cost table used for cost planning shows cost per production line, which is manufacturing costs broken down into direct labor costs and indirect line costs. In standard costing, shops or cost centers are groups of two or more lines.

In productivity analysis, the major steps in the production process of the new subassembly are analyzed, and the sum of their costs is compared to the target cost of the subassembly. If the expected cost is too high, the section leaders responsible for each step in the production process are asked to identify a cost-reduction target for each step.

At Komatsu, ultimate responsibility for these cost-reduction targets lies with the product manager, who is responsible for ensuring that the new product successfully enters production. If the initial aggregated cost reductions are insufficient to allow the subassembly to be manufactured for its target cost, then the product manager and the production staff negotiate to increase the expected productivity savings. The final aggregation of the negotiated cost-reduction targets when subtracted from the current cost provides the latest estimate of the target cost.

The process of productivity analysis at Komatsu is illustrated by the redesign of a mounting socket in the main frames of the firm's bulldozers. In the old design, the mounting socket consisted of a hole drilled through the body of the frame. This design was simple to manufacture but had the drawback of creating a stress zone around the hole. To ensure that the mounting socket was strong enough, that section of the main frame had to be manufactured from expensive, high-grade materials. Productivity analysis identified lowering the amount of high-grade material in the main frame as one way to reduce costs. The new design involved welding a mounting bracket containing the mounting socket hole to the vehicle main frame. The new mounting unit was designed to reduce the strain imposed on the frame so that normal-grade steel could be used.

Reducing the Workload

When the number of components in products is especially large, techniques to reduce the cost of developing target costs at the component level have to be developed. Firms may estimate target costs for similar families of components from a base case. For example, at Nissan, to avoid having to develop target costs for all 20,000 components in a typical new model line, the engineers perform detailed target costing on only two or three representative variations. Each variation contains approximately 3,500 components, and typically 80% of the components are common across variations, so about 5,000 components are subjected to detailed target costing. The target costs of the other 15,000 components are estimated by comparing them to similar components in the 5,000 already target costed. This target costing exercise provides cost-reduction objectives for all the components in the new model.

Applying Supplier-Base Objectives

The supplier-base objectives include maintaining supplier relations, extending the supplier base, and inducing supplier creativity. Target costing is most effective when applied in the context of long-term supplier relations that are cooperative in nature. The objective is to create a supply chain culture of continuous innovation and cost reduction, but it can be achieved only if the suppliers view the buyer-supplier relationship as worth the investment. Supplier reputation for innovation plays an important role in determining the willingness of the firm to accept slightly higher prices or occasional lower levels of innovation and still grant some of the business. The objective is to ensure that innovative suppliers are retained in the supplier base. For example, at Isuzu although the supplier rated as having the highest value generally will win the order, firms that have a reputation for being good suppliers often are awarded at least part of the order even if their products do not have the highest value. Examples of such companies include Yuasa for batteries, Toyo Valve for valves, and Nihon Seiko for bearings. These firms are awarded partial contracts to maintain their relations with Isuzu.

To increase the rate of innovation and in particular to enable new technologies and production processes to be adopted, the firm must continuously look for new suppliers. The objective is to identify suppliers that are highly creative and innovative or have developed considerable expertise in technologies that the firm is now incorporating into its products. For example, if a firm shifts from using mechanical to electronic controls in its products it must identify a new set of suppliers that have expertise in electronic control systems.

Finally, many firms use incentive plans to encourage their suppliers, both to reward innovation and to signal where additional cost reduction should occur. These plans reward the supplier with all or part of the order for a given component. Nissan uses such an incentive plan to motivate its suppliers. For example, if a cost-reduction idea is accepted, the supplier that suggested the idea is awarded a significant percentage of the contract for that component for a specified time period, say 50% for 12 months. This incentive scheme is viewed as particularly important because even if a cost-reduction objective cannot be achieved for this model, it signals to the suppliers that when the next model is developed, this component will be subject to cost-reduction pressures.

The primary benefit of this "maximize value" approach is its ability to bring out the strengths of each supplier. Where ethical, Isuzu shares innovations made at one supplier with other suppliers to help them achieve target costs and make an adequate return. Originality also plays a role in the selection of the winning design. If one of the suppliers finds a way to add additional functionality to the component that increases its value, Isuzu engineers incorporate this functionality into the part's specifications. Typically, the creative supplier achieves a higher value than the other suppliers because its component already contains the extra functionality. In this way, suppliers are encouraged to act as ancillary research and development laboratories for Isuzu. Incorporating the additional functionality spreads the innovation to other suppliers, thus increasing their ability to provide higher-value-added components in the future.

Selecting Suppliers

One of the critical decisions taken during the component-level section of the target costing process is the sourcing of components. Suppliers are selected based on three criteria: the competitiveness of their bids, their reputation, and the degree of innovation they have brought to the component. The bids are taken as early as possible in the target costing process and are incorporated by an iterative process into the component-level target costs. This process is designed to ensure that the individual component-level target costs are achievable but sum in total to the component-level target cost of the product.

Supplier reputation comes into play because of the firm's desire to maintain relations with suppliers that have excellent reputations. Therefore, when a long-term supplier fails to make the lowest bid or develop the most innovative solution, the firm may still award it part of the contract. For a given component, the degree of innovation a supplier introduces influences the value the firm associates with the component. The higher the degree of innovation, the higher the value, all else being equal. Since the firm wants to reward innovation, it will usually select the most innovative design.

Supplier Negotiations

The bidding process is critical because the target costs for externally acquired components are typically set through negotiation. This process starts with suppliers (both internal and external) providing estimates of their selling prices to the company. At Nissan the first step in the product development stage is to prepare a detailed order sheet for the new model. This order sheet lists all the components required in the new model. It is analyzed to see which components are likely be sourced internally and which externally. Both internal and external suppliers are given a description of each component and its potential production volumes. Suppliers are expected to provide price and delivery-timing estimates for each component.

The comparison of the target cost of each function and the sum of the expected cost of the components in that function after cost reduction indicates whether each major function can be produced at its target cost. When the sum of the component costs is too high, additional cost reductions are identified until the total target cost of the one expected to sell the highest volume is acceptable. The target costs for each component are compared to the prices quoted by the suppliers. If the quoted prices are satisfactory, the quote is accepted. If the initial quote is too high, further negotiations are undertaken until an agreement is reached.

These negotiations are iterative in nature and are a powerful mechanism for cost management. The nature of the negotiations depends on the relative strength of the buyer-supplier relationship. Isuzu provides capital and design support to some suppliers so they can produce the advanced components required for the next generation of products. For other firms, the relationship is less important, and the supplier offering the greatest value is chosen. For these types of bids, three suppliers are contacted for each component and asked to develop prototypes. The suppliers are told the target quality, functionality, and price of the component and are expected to produce prototypes that satisfy all three requirements. Once the three prototypes are submitted, Isuzu engineers analyze them to determine which one provides the best value. The product with the highest perceived value is selected, and the supplier is awarded the total contract.

The nature of the negotiations also depends on the degree of reliance the buyer places on the supplier for research and development.

At Komatsu, though target costs are supposed to be negotiated with suppliers, management was concerned that in reality these negotiations were relatively one-sided, with Komatsu dominating. In addition, management felt that the suppliers were brought into the negotiations too late in the design process. To allow the suppliers greater input into the design process, Komatsu initiated periodic meetings between the suppliers' research and development staffs and its own. The aim of these meetings is to integrate the research and development efforts of the two groups, allow suppliers to provide input much earlier in the design process, and ensure more substantive target cost negotiations.

To maintain the discipline of target costing, target costs of components can be relaxed only under very special circumstances and typically only for a short period of time. The objective of this relaxation is to allow the supplier to make a reasonable return while finding ways to reduce costs to the target levels.

Applying the Cardinal Rule

The completion of the target cost setting process for components signals the achievement of a major step in the product design process. The anticipated cost of the product can now be determined by summing the costs of all the components, group components, and major functions either produced internally or acquired externally. The sum of all these costs has to be equal to or lower than the target cost of the product; otherwise redesign is required. At Olympus, once a new product has passed the research and development design review, it is released to Tatsuno production for evaluation. The design review determines where and how the new product will be produced. To aid these decisions, a detailed production blueprint is developed. It identifies both the technology required to produce the camera and the components it contains. Using this blueprint and cost estimates from suppliers and subsidiary plants, the chief engineer reestimates the production cost of the product. If this cost is less than or equal to the target cost, the product is submitted to the division manager for approval for release to production.

Thus, the cardinal rule continues to operate throughout the design process. The interaction between design and manufacturing is critical to ensure that the new product's manufacturing cost is indeed its target cost. Without such interaction, the target cost and manufacturing cost could be significantly different, rendering the target costing system ineffective.

SUMMARY

The third section of target costing consists of decomposing a product's target cost to the component level. The objective of this stage is to transmit the cost pressure faced by the firm to its suppliers. The target costs set for externally acquired items become the suppliers' selling prices. Thus, an outcome of target costing is that the buyer sets the selling prices for its suppliers. In many cases supplier inputs are used to help set the target costs to ensure their achievability. Just as in setting the profit plans and product-level target costs, it is critical to the success of the target costing system that the target costs set for suppliers be achievable most of the time.

The decomposition of product-level target costs into component-level ones consists of two major steps. The first is to decompose the target cost from the product to the major function level; the second is to decompose major function-level target costs to the component level (see Figure 8-15). The chief engineer controls the process of setting the target costs of major functions and establishes the main themes of the new product. These themes identify the primary functional improvements that differentiate this model from its predecessor. The chief engineer decides on the target costs of each major function after discussions with the various design teams responsible for the major functions. At most companies these teams report both to the chief engineer and to their division heads. This matrix structure ensures that the designs of the major functions

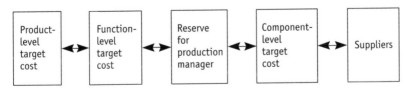

FIGURE 8-15. COMPONENT-LEVEL TARGET COSTING

reflect the demands of the chief engineer and share a common design philosophy across products. The cardinal rule of target costing is enforced by ensuring that the target costs of the major functions sum to the target cost of the product.

In the second step, the design teams decompose the target costs of the major functions down to the component level. The way component-level target costs are set depends on whether the component is designed by the firm or a supplier. If the component is designed by the firm, its engineers have considerable knowledge of the manufacturing process and can identify ways to reduce subcontractors' costs. If, in contrast, it is designed by the supplier, then typically the firm's engineers have less knowledge of the manufacturing process and can only identify the target cost but not suggest ways to achieve it. The chief engineer typically takes little part in this decomposition but sometimes will specify the target costs for "main theme" components or especially expensive ones. The cardinal rule of target costing is enforced by ensuring that the target costs of the components sum to the target costs of the major functions that contain them.

A critical decision taken at this stage of the product development process is the selection of suppliers. Supplier selection is critical because the component-level target costs are set in part via a negotiation process. By carefully managing the selection of suppliers and transmitting the competitive pressure to them, the firm can induce suppliers to focus their creativity on finding ways to reduce costs and increase the functionality of their products. This pressure is reinforced by incentive schemes that reward supplier creativity.

Target costing thus aligns supplier creativity with that of the firm's product designers, achieving the product functionality, quality, and price the customer desires at a cost that allows the planned levels of profit to be earned. While the three sections of target costing—setting the allowable cost, product-level target costing, and component-level target costing—are present in all target costing systems, their relative importance and the way they are accomplished varies by firm. Chapter 9 explores the factors that influence the target costing process.

FACTORS THAT INFLUENCE
THE TARGET COSTING PROCESS

INTRODUCTION

A firm's strategy for a given product is reflected in the way the product's selling price, quality, and functionality change over time. For some firms, such as Toyota, Nissan, and Komatsu, the strategy is to increase functionality as quickly as possible while essentially holding selling prices constant. For others, such as Olympus and Sony, it is to increase functionality while allowing selling prices to fall. At Topcon, the ability to add functionality often allows the firm to raise the selling price.

A firm's competitive strategy strongly influences the way the interaction among the three characteristics of the survival triplet is handled in the target costing process. The primary purpose of Nissan's and Toyota's target costing systems is to achieve the target cost by creating downward pressure on suppliers' selling prices and by modifying the functionality of the product. Komatsu sets the product's target selling price and then determines the functionality that can be supported at that price. Komatsu's target costing system focuses primarily on cost control, not functionality management.

Olympus determines the distinctive functionality of its products first and then sets target selling prices. In subsequent steps, the firm determines the other functional characteristics that the target selling prices can support. Unlike Nissan, which can reduce functionality right up to the launch date by simply omitting features, Olympus must set product functionality before manufacturing commences. Consequently, setting the functionality of the product is central to the focus of Olympus' target costing system.

The type of competition these firms face plays a role in determining at what point in the development cycle the selling price of a product can be decided. At Nissan, Toyota, and Komatsu, competition is based primarily on keeping prices constant and increasing product functionality with each new generation, so selling prices can be decided fairly early in the development cycle. In contrast, the selling prices of Topcon's products are determined in part by their overall functionality. The sensitivity of the target selling price to the product's functionality makes it difficult for Topcon to set a target selling price until the functionality of the product is known.

Despite the different relationships between price and functionality, all the firms use target costing systems that have the same generic structure. This structure consists of three major subprocesses: market-driven costing, product-level target costing, and component-level target costing. Even though the underlying structure is identical, there are clear differences among the six target costing systems described in the cases (see Part III). These differences are driven by numerous factors, but five appear to play a critical role in shaping the way the firm approaches target costing. Of these five factors, two predominantly shape the market-driven target costing section of the target costing process, two the product-level target costing section, and one the component-level target costing section. Some of the factors influence more than one section of the target costing process. These multisection effects will be discussed after the primary effects.

The three sections of target costing are obviously not independent. In general, as the benefits from one section decrease the overall energy expended on target costing in the other sections will also decrease. However, sometimes this relationship is obscured by the difference between the ease of accessing the information required by a section in the target costing process and the value of applying

Value of Information

	Low	High
High **Ease of information collection** **Low**	Little energy expended on target costing	Moderate energy expended on target costing
	No energy expended on target costing	Considerable energy expended on target costing

FIGURE 9-1. THE INTERACTION OF EASE OF INFORMATION
COLLECTION AND THE VALUE OF THE INFORMATION

the technique it relies on (see Figure 9-1). For example, if customers have similar future requirements and can explain them clearly, then little customer analysis is required even though it is valuable. In contrast, if customers have diverse future requirements and have very little understanding of them, then the customer analysis will be time consuming. It will be of little value and may not be worth the effort.

FACTORS THAT INFLUENCE MARKET-DRIVEN COSTING

The factors that help shape the market-driven costing section of the target costing process are the *intensity of competition* and the *nature of the customer*. These two factors help determine the nature and extent of the information collected about customers and competitors in the market analysis process. They also help determine how difficult it will be to ensure that products are inside their survival zones when launched and hence how much benefit target costing will bring.

Intensity of Competition

It is reasonable to suspect that the intensity of competition is a factor to consider since it has been shown in other environments to

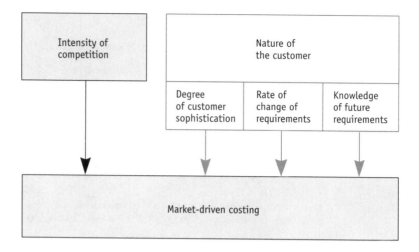

FIGURE 9-2. FACTORS THAT INFLUENCE MARKET-DRIVEN
COSTING: INTENSITY OF COMPETITION

influence the energy expended on cost management.[1] The intensity of competition influences how much attention the firm should pay to competitive offerings in the target costing process (see Figure 9-2). All the firms studied could identify four to six direct competitors who technologically were fairly evenly matched. These firms had adopted a confrontational strategy because they couldn't develop sustainable competitive advantages over each other. Characteristics of confrontational markets are low profit margins, low customer loyalty, low first-mover advantages, and dramatic failure of products launched outside their survival zones.

Under such conditions, the benefits of the market-driven portion of target costing are high. Confrontational firms typically expend considerable effort on competitive analysis to try to estimate what future competitive offerings will look like. In addition, they frequently purchase their competitors' new products and use teardown methods to see what they can learn from them. The low profit margins and lack of customer loyalty mean that a firm cannot afford to make too many mistakes when launching new products.

---

[1] Khandwalla, P. N., "The Effect of Different Types of Competition on the Use of Management Controls," *Journal of Accounting Research*, Autumn 1972, pp. 275-285.

By transmitting the competitive pressure faced by the firm to its product designers and suppliers, target costing increases the probability that new products will be inside their survival zones when launched. In contrast, when competition is less intense, nonconfrontational strategies such as cost leadership and differentiation can be successful. Such strategies allow for higher profits and increased customer loyalty, and the benefits of target costing will be lower in such environments.

Target costing is particularly valuable for firms that have adopted confrontational strategies because failure to launch products that are in their survival zones typically leads to rapid and significant loss of market share. These losses are driven by the narrow survival zones that result from equivalent competitors chasing the same customers.competitors' ability to rapidly bring out me-too products makes it difficult for firms to recoup their investments in product development. The rapid copying leads to shorter life cycles, and the inability to reap first-mover advantages leads to lower profits.

Thus, the firm is forced to amortize its development costs over fewer units that are generating lower profits. The ability of the successful products to offset failures is reduced, putting significant pressure on the firm to minimize product failures. Therefore, it is conjectured that as the intensity of competition increases, so does the value of target costing to the firm. For example, Sony has managed to differentiate its products based on their superior functionality over those of its competitors. This lowered intensity of competition may be one of the main reasons Sony has a less well-developed target costing process compared to many other firms in our sample. In contrast, all the other firms are in confrontation and, with the exception of Topcon, have well-developed and extensive target costing systems.

Nature of the Customer[2]

Many characteristics of customers can influence the intensity of consumer analysis the firm should undertake, but evidence suggests

---

[2] In this volume, we are exploring target costing practices of firms where the buyer (i.e., customer) has little or no individual power over the supplier firm. The next volume in the series, on interorganizational cost management techniques, will explore chained target costing where the buyer wields considerable power.

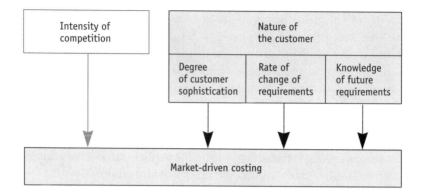

FIGURE 9-3. FACTORS THAT INFLUENCE MARKET-DRIVEN
COSTING: NATURE OF THE CUSTOMER

that three are particularly important in helping determine the benefits of target costing, in particular the market-driven costing section (see Figure 9-3): the *degree of customer sophistication*, the *rate at which future customer requirements are changing*, and the *degree to which customers understand their future product requirements*.

These three characteristics help determine the benefits a firm can potentially derive from target costing because they deal with the shape, rate of change of location, and ease of predicting the location of survival zones. Analysis of the practices observed in the six companies suggests that target costing is particularly valuable for firms that have to compete in environments with narrow survival zones that change location rapidly but are relatively predictable.

Degree of Customer Sophistication

The degree of customer sophistication determines how good customers are at detecting differences between the price, quality, and functionality of competitive products. Sophisticated customers are highly educated about available product offerings, can detect minor differences, and will switch freely among manufacturers to buy the best products. As customers become more sophisticated, the survival zones of products become narrower. When survival zones are narrow, it is easier to launch products that fall outside them and hence fail. To increase the probability of launching products inside these narrow survival zones, firms expend considerable energy on consumer analysis, trying to determine where survival zones will be when the product is launched.

In the automobile industry, the primary characteristic of the survival triplet used to differentiate products is functionality. Firms compete by continuously increasing the functionality of their products while keeping the price and quality essentially unchanged. Customers, therefore, have come to expect a steady rise in product functionality and have quite clear expectations for their future purchases. For example, to ensure the success of their products, Toyota and Nissan both do a great deal of consumer analysis to identify future products that will both satisfy their customers and sell sufficient volume to be profitable.

The same holds true in the camera industry, where most consumers are highly sophisticated and capable of identifying the exact features they expect in a new camera. Survival zones are very narrow in that industry; there is no price freedom. Olympus collects qualitative information about consumer preferences and trends from seven sources, including recent purchases, professional photographers, and focus groups. In addition, the firm monitors its competitors' actions closely. Sources of competitive information include press and competitor announcements, patent filings, and articles in patent publications. The firm uses this information to predict what types of products competitors will introduce in the short and long term and what their marketing plans are.

Such intense analysis of both customers and competitors is necessary because of the high price of failure to launch products in their survival zones when customers are highly sophisticated. For example, in the mid-1980s, Olympus' camera business began to lose money, and by 1987, its losses were considerable. Top management ascribed these losses to a number of internal and external causes. The major internal causes were poor product planning, a lack of "hit" products, and some quality problems. While Olympus' overall quality levels were above average for the industry, certain products that relied on completely new technologies had rather high defect rates. These quality problems had caused Olympus' reputation to suffer. The problem was simple—Olympus' sophisticated customers had rapidly found that the firm's cameras were inferior to its competitors' and switched accordingly.

Target costing becomes especially valuable in environments with highly sophisticated customers because survival zones are narrow and therefore products must be designed that satisfy customer requirements as closely as possible. Without the discipline of target costing engineers sometimes add extra functionality to products in the belief that it will make them attractive to customers. Unfortunately, these extra features often cost more than the value the customer places on them. The outcome of such design "improvements" is products that cost too much and have profits below expectations. In confrontational environments profits are already low and there is little room for error, so the discipline target costing places on the product designers is critical to a firm's survival.

In environments with sophisticated customers, the target costing process will have a strong external orientation because understanding the customers' requirements is critical. In contrast, in environments where consumers are less sophisticated, target costing will not be as beneficial and will be more internally focused.

The Rate at Which Customer Requirements Change

The rate at which customer requirements change defines how quickly survival zones move over time. When survival zones are moving rapidly, it becomes more difficult for a firm to predict where a new product's survival zone will be when it is launched and to ensure that it will be inside the zone. In the automobile industry customer expectations change relatively quickly and so Nissan samples consumer preferences on a regular basis during the product design process. For example, the market is sampled when the product is first conceptualized, just before it enters the product design stage, and just before it enters the production stage. The primary purpose of these market revisits is to capture changes in the position of survival zones since the last survey. The product's design is then modified when possible to increase its probability of success.

In contrast, Komatsu's customers are commercial buyers, not consumers. They are highly sophisticated and well aware of their preferences. Given the nature of the firm's products—bulldozers and excavators do not change rapidly—Komatsu's target costing system expends considerably less energy on customer analysis than Nissan's or Toyota's. It's much easier for Komatsu to keep track of changing customer expectations than it is for an automotive company.

Target costing is more beneficial in environments where consumer preferences change rapidly because under such conditions a firm is more likely to launch products that are outside their survival zones. Firms with such customers must expend considerable effort on predicting future customer requirements. In contrast, when customer requirements are stable, less effort is required to locate the position of a product's survival mode, and target costing provides smaller benefits. Reflecting the diminished benefits, the target costing systems at these firms are more internally focused.

Degree of Understanding of Future Product Requirements

The degree of understanding of future requirements in part determines how much customer analysis is done in the target costing process. As the degree of understanding increases, it becomes more beneficial to rely on known customer preferences to determine the future location of survival zones. In contrast, when customers have little understanding of their future requirements, firms that pay too much attention to customers risk launching products that fail because they are outside their survival zones.

In the earth-moving business, customers have a high degree of awareness of their future requirements. Komatsu can rely on its customers to tell the firm what needs to be improved in their designs and to a certain extent by how much. The customers have a clear idea of their future requirements. In such an environment, target costing will offer considerable benefits because the customer is able to specify quite accurately the location of future survival zones.

In contrast, in the consumer electronics industry consumers have a lower degree of understanding of their future requirements, and product failures are more common. For example, at Sony not all attempts to proliferate the Walkman line were successful. There were several notable failures, including the dual cassette model and the rewinding headphone model. The dual cassette model let the user record from one tape onto another, and it was necessarily thicker than existing models. Consumers reacted negatively to this model, and very few were ever sold. Consumers had the same reaction to the automatically rewinding headphone model; while it had the added advantage of automatically retracting the headphone wire when the user was finished listening, it, too, was much thicker than other models. The thickness of a Walkman appeared to be a critical attribute. The critical attribute often becomes apparent only after the firm has launched a new product. Consequently, target costing is less beneficial in environments where the future locations of survival zones are hard to predict.

FACTORS THAT INFLUENCE PRODUCT-LEVEL TARGET COSTING

The factors that help shape the product-level target costing section of the target costing process are the firm's *product strategy* and the *characteristics of the product*. These two factors help determine the nature and extent of the information collected about historical cost trends and customer requirements. The product strategy establishes the number of products in the line, the frequency of redesign, and the degree of innovation in each generation of products. The characteristics of the product include the complexity of the product, the magnitude of the up-front investments, and the duration of the product design process.

Product Strategy

The evidence suggests that product strategy is a primary determinant of the degree of effort expended on product-level target costing and where and how that effort is expended. Three characteristics of a firm's product strategy help determine the benefits that will be derived from product-level target costing. They are the *number of products in the line*, the *frequency of redesign*, and the *degree of innovation* (see Figure 9-4).

Number of Products in the Line

The number of products in the line[3] is a major determinant of the total product development budget. As the number of products increases, so do the product development budget and the benefits

---

[3] Number of products excludes minor variations such as color.

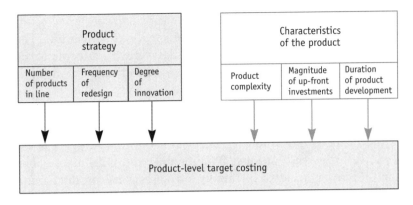

FIGURE 9-4. FACTORS THAT INFLUENCE PRODUCT-LEVEL
TARGET COSTING: PRODUCT STRATEGY

derived from product-level target costing, all else being equal. The benefits of target costing increase because more money is involved, and it is worth expending greater energy to reduce risk.

The number of products in the line must be carefully managed. Differing customer requirements can be satisfied by developing products differentiated either vertically or horizontally. Vertically differentiated products differ by the degree of functionality they provide and their selling price. The higher the price, the higher the functionality (and perhaps quality) of the product. Horizontally differentiated products sell at the same price but deliver a different bundle of quality and functionality. Relatively small variations in functionality and price are often achieved by developing optional features such as a Corolla automobile with or without a passenger air bag. In contrast, major variations in functionality are achieved by introducing different product models, for example, a Corolla versus a Camry.

The greater number of different products a firm supports, the higher the overall level of customer satisfaction. The evidence suggests that as the number of products in the line increases, so does the effort expended on target costing because new product launches occur more frequently. This observation is intuitively reasonable; because target costing operates predominantly at the individual product level, the benefits must occur at that level. For example, Olympus had a relatively ineffective target costing system prior to the reconstruction of their camera business. As part of their strategy to reconstruct their camera business, they significantly increased the number of horizontally differentiated products in their line. The enhanced benefits from target costing that were the outcome of the increased number of products might have helped motivate the decision to upgrade the firm's target costing system.

Frequency of Redesign

At the heart of the product strategies of the sample firms is the objective to increase product functionality as rapidly as possible. This objective is achieved by rapidly introducing new products, with each new generation incorporating the latest technology and providing increased functionality. In all the firms, product development times have been reduced so as to introduce new products more frequently. At Olympus, rapid introduction is considered important because it allows the firm to react in a timely fashion to changes in the competitive environment. One of the key elements in improving the firm's ability to react quickly was reducing to 18 months the time required to bring new compact cameras to market. The equivalent benchmark when the OM10 SLR camera was developed in 1980 was 10 years.

Intense competition has forced the firms to become experts at developing and launching products at a rapid rate. However, this ability has a downside. First, the short manufacturing phase implies lower sales volumes for each product, which means that the time available to generate an adequate return on the up-front investment is limited. To remain profitable, firms must launch a high percentage of profitable (as opposed to unprofitable) products. Second, due to the short product life cycles, there is not enough time to correct any errors. If an unprofitable product is launched, it often will remain unprofitable until it is withdrawn; it becomes critical to design new products so that they are profitable.

Consequently, the higher the rate of product introduction, the greater the benefits derived from target costing because the product development budget is higher, and therefore more is at risk. Such firms are expected to have well-developed target costing systems that subject the design process of all new products to systematic cost-reduction pressures. In contrast, firms that rarely introduce new products will not require formal target costing systems, but they will probably apply target costing principles on an ad hoc basis as required.

Degree of Innovation

The degree of innovation in each new product generation helps determine both the magnitude of the product development budget and to what extent historical cost information can be used to estimate future costs. As the degree of innovation increases, so does the cost of product development. In addition, information about past products becomes less valuable. Historical cost information about earlier products will have little value, especially for revolutionary products that rely on completely new technologies. Similarly, customer, competitor, and supplier information can be invalidated by significant innovations in product design. In contrast, for products similar to the ones they are replacing, the past is often highly predictive of the future, and value engineering techniques, such as functional analysis, that depend on the use of the same technology can be applied.

Target costing is most difficult to apply to revolutionary products. First, target selling prices are often difficult to establish because the value to the customer of the new product is difficult to estimate. Also, because the firm has never applied the technology in its products, historical cost information is not available. Finally, more new suppliers are typically involved. When the new model does not rely heavily on existing designs, the target costing system is of less value; more intuition is required, as opposed to hard facts. For example, when Toyota introduced the Lexus, they derived less benefit from target costing because of the new vehicle's high degree of innovation.

At Toyota, there are two broad categories of product development. The first is for completely new types of automobiles, and the other is for changes to existing models. The development of an entirely new model such as the Lexus is relatively unusual; projects to modify existing models are normal. Toyota uses target costing primarily to support model changes, although the same general cost control procedures are used for the design of entirely new vehicles. The primary difference between the procedures used for the two types of projects is the level of uncertainty in the cost estimates, which are much higher for projects involving new types of automobiles.

Characteristics of the Product

Three characteristics of the product have a particularly strong influence on the benefits derived from target costing and the way it is practiced. These characteristics are the *product complexity*, the *magnitude of up-front investments*, and the *duration of the product development process* (see Figure 9-5). The complexity of the product affects how difficult it is to manage the product design process. The up-front investment involves the amount of capital consumed in the research and development process, getting ready for production, and actually launching the product. The duration of the product development process captures the time it takes to go from product conception to release to production.

Product Complexity

Product complexity captures the number of components in the product and the number of distinct production steps required to manufacture it, the difficulty of manufacturing the components it contains, and the range of technologies required to produce them. As the complexity of the product grows, the benefits of target costing in-

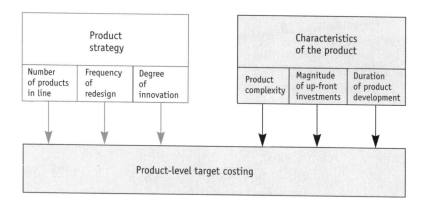

FIGURE 9-5. FACTORS THAT INFLUENCE PRODUCT-LEVEL TARGET COSTING: CHARACTERISTICS OF THE PRODUCT

crease for two major reasons. First, the degree to which costs can be influenced in the product design stage versus the manufacturing stage increases. Second, it becomes more difficult to manage the product design process and ensure that component-level target costs sum to the product-level target cost. Therefore, the benefits of target costing are expected to increase with the complexity of the product.

However, as the complexity increases, so does the cost of applying target costing at the component level. Fortunately, there are ways to simplify the target costing process to reduce the effect of product complexity by performing detailed target costing on only two or three representative variations, as opposed to all of them. Consequently, as product complexity increases, target costing becomes more beneficial, and ways to reduce the costs of performing target costing emerge. Toyota, Nissan, and Komatsu manufacture products that are considerably more complex than the other firms' products. Their target costing processes reflect this increased complexity by being more formalized. This formalization helps the firms cope with the large number of components that have to be subjected to target costing.

Magnitude of Up-Front Investments

Firms that produce products with very low product development costs (such as soft drinks) are often willing to launch numerous products each year with the expectation that only a few will be successful. Thus, when up-front investments are small, the benefits of target costing are lower. In contrast, as the size of the up-front investment increases, the number of products a firm is willing to launch typically will decrease because the firm will be less willing to take risks. Firms that produce products with high up-front investments usually develop a fairly small range of products, each carefully designed to satisfy a specific market segment. For example, at Nissan in recent years, consumer analysis identified more than 50 potential models that Nissan theoretically could introduce successfully. However, top management has identified the optimum number of models that Nissan can successfully support at under 30. This number is limited by several factors, including the cost of differentiating each model in the minds of consumers, the investment in research and development, and the cash flow associated with maintaining dealer floor inventory. Thus, the challenge that Nissan management faces is to select the approximately 30 models that will maximize market coverage.

The benefits of target costing will be greater for firms that have products with high up-front investments because every product has to have the maximum probability of being successful. Target costing is even more important for firms with products with high up-front investments and short manufacturing lives, because it is critical that any products launched have adequate profit levels and sales volumes. Under such conditions, careful product selection is critical, and target costing can play an important role in helping ensure that product profitability is adequate.

Finally, for high up-front investment products, life-cycle analyses are especially important. Therefore, life-cycle target costing is more commonly practiced in such firms than in those producing products with low up-front costs. The profits earned by the product must be sufficient to pay back the heavy investment in product development, and the target profit margins must reflect the level of investment. For this reason, Nissan uses life-cycle analysis to justify the launching of new automobiles whereas Sony does not.

Duration of Product Development

The length of time it takes to develop a new product also helps determine the benefits derived from product-level target costing. As the duration of design gets longer, the probability that the conditions used to validate the design of the new product might change increases. Therefore, for products with long development cycles, such as automobiles and bulldozers, the target costing system needs to contain several stages at which all pertinent conditions are reviewed. In contrast, for products with short development cycles, such as cameras and consumer electronics, fewer reviews are required. Thus, as the product design cycle increases in length, the target costing system typically becomes more complex.

The product development cycle for automobiles, six years, is relatively long. This extensive period requires multiple reviews of market conditions and decision points about continuing the project. At Nissan and Toyota, reviews occur at the beginning and end of the conceptual design stage and during the product design stage. Just prior to entering production, the firms make a final adjustment to the specifications of the new model to make sure that it achieves its target cost. Just prior to product launch, the firms decide exactly which features will be treated as optional versus standard. This fine-tuning ensures that the target cost will be achieved when possible and that the new model will satisfy the customer.

The longer product development cycles make target costing more beneficial because the long time between design and launch increases the risk that new products will be unsuccessful. In addition, longer product development cycles typically lead to more formal target costing systems with multiple decision points, reflecting a more disciplined product development process. Even when product development time is short, as is the case with Olympus cameras, target costing does not appear to introduce any significant delays into the process. The target costing process is so integrated into the market analysis and product development processes that most, if not all, the extra work required by the target costing process can be undertaken in parallel.

FACTOR THAT INFLUENCES COMPONENT-LEVEL TARGET COSTING

The factor that influences the component-level target costing section of the process is the firm's *supplier-base strategy*. This strategy helps determine the benefits that can be derived from component-level target costing because it shapes the amount of information the firm has about the costs and design capabilities of its suppliers.

Supplier-Base Strategy

Three aspects of the supplier-base strategy have a particularly strong influence on the benefits derived from component-level target costing. These characteristics are the *degree of horizontal integration*, which captures the percentage of the total cost of the firm's products that are sourced externally; the *power over suppliers*, which helps establish the ability of the firm to legislate selling prices to its suppliers; and the *nature of supplier relations*, which deals with the degree of cooperation the firm can expect from its suppliers and in particular the amount of design and cost information sharing (see Figure 9-6).

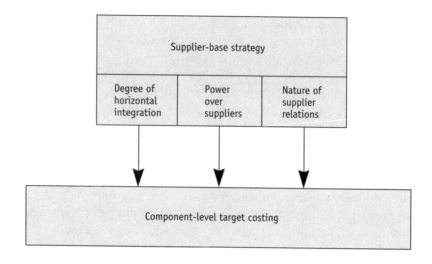

FIGURE 9-6. FACTOR THAT INFLUENCES
COMPONENT-LEVEL TARGET COSTING

Degree of Horizontal Integration

Lean enterprises typically are horizontally, not vertically, integrated. They buy a large part of the inputs required to produce their products from external sources. The higher reliance that lean enterprises place on external suppliers increases the importance of supplier management and hence, component-level target costing. Two primary factors increase the potential benefits of component-level target costing. First, since a greater percentage of the product is externally sourced, the potential savings are greater because target costs can be developed for each of the externally acquired components and can be used to help create pressure on suppliers to reduce their prices. In contrast, in vertically integrated firms it is often difficult to put effective pressure on the other divisions to reduce their costs. Second, the returns from focusing supplier creativity are greater. Suppliers not only provide a higher percentage of the firm's products, they are also responsible for a greater part of the design. For example, Komatsu's suppliers are asked to design and produce complete engine cooling systems instead of producing individual components such as radiators, electric motors, and fans. Therefore, creating incentives for Komatsu's suppliers to be more innovative in their designs generates greater payback.

Power Over Major Suppliers

The relative power of buyer-supplier relations determines how much energy a firm puts into determining purchase prices for components. When buyer power is high, buyers will frequently spend considerable effort developing component-level target costs (i.e., purchase prices) for purchased components. In contrast, firms with low production volumes and little buyer power will expend less energy on developing target costs for purchased components because suppliers will not accept them as the selling prices for their products (unless they provide adequate returns).

The more power the firm has over its suppliers, the more benefits it can derive from target costing by using it to create cost pressures on its suppliers. In contrast, when a firm has little power over its suppliers, the benefits of target costing will be reduced. For example, Topcon has little power over its suppliers due to the low volume of specialty ophthalmic equipment it sells, and therefore it does very little to develop component-level target costs. In contrast, the other firms have considerable power over their suppliers and have sophisticated component-level target costing systems.

Nature of Supplier Relations

The evidence suggests that as supplier relations become more cooperative, the target costing process in general and the component-level step, in particular, become richer and more beneficial. At the heart of the increased benefits lies the ability of the buyer and supplier firms to combine their design creativity to find better ways to reduce costs. For example, Komatsu's design engineers frequently visit their suppliers and help them with design problems.

This cooperation can be supplemented by a number of interorganizational cost-management techniques.[4] By means of these other mechanisms product designers and suppliers can have joint meetings and frequent interactions and pool their expertise to find creative solutions that increase functionality and quality or reduce costs. When used in this cooperative setting, component-level target costing still places suppliers under considerable cost pressure,

---

[4] See the second volume in this series.

although this pressure is offset to some extent by the product designers helping suppliers to find ways to achieve their cost-reduction objectives. In contrast, in adversarial supplier relations, component-level target costing can be used to force selling prices on the firm's suppliers, but there is no mechanism to take advantage of any synergy between the designers of the two firms.

How the Factors Influence the Target Costing Process

The foregoing factors influence the target costing process in six ways (see Figure 9-7). Some of them, such as the intensity of competition and the sophistication of customers, alter the width of survival zones.

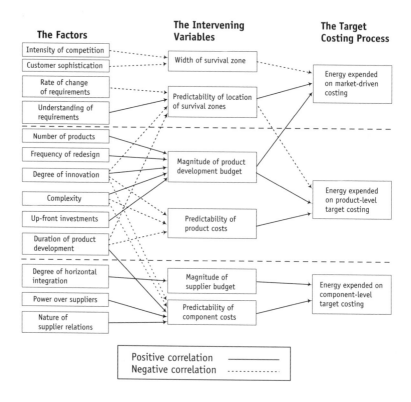

FIGURE 9-7. How the Factors Influence
the Target Costing Process

As survival zones narrow, it becomes more difficult for the firm to launch successful products, and therefore more energy is expended on market-driven costing. Other factors, such as the rate of change of customer requirements and the degree of understanding of future requirements, affect how easy it is to predict the location of survival zones. The quicker survival zones move, the more difficult it is to predict their future location and the greater the difficulty the firm will have in launching successful products, so, again, more energy is expended on market-driven costing. Thus the evidence suggests that the benefits of target costing increase in such environments because the greater coordination between marketing and product design increases the likelihood that the product will satisfy the customers.

Other factors, such as the number of products, frequency of redesign, degree of innovation, product complexity, and size of up-front investments, help determine the product development budget. The budget can be increased by adding additional products to the line, accelerating the frequency of redesign, and developing products with a greater up-front investment. As this budget increases, so does the value of product-level target costing, because more capital is at risk.

The degree of innovation, the complexity of the product, and the duration of product development all influence the ease of predictability of the cost of the new product. While much of the cost of the products produced by lean enterprises is due to externally sourced components, the cost of the product also includes the internally sourced components, the assembly costs, and other production costs. These all have to be estimated when undertaking product-level target costing.

As the degree of horizontal integration increases, so does the supplier budget as more of the product is outsourced. The higher the degree of outsourcing, the greater the benefit that can be derived from component-level target costing because more is at risk.

Finally, some of the factors influence the ease of predicting future component costs. As the degree of horizontal integration increases, the importance of component-level target costing increases because a higher percentage of the product is externally sourced. Next, the power over suppliers is important because it controls the degree to which the firm can dictate purchase prices to its suppliers using its target costing system. Finally, the more cooperative the supplier relations, the greater the opportunity to take advantage of any synergy between the firm's design engineers and those of the suppliers. Thus, all these factors influence the component-level target costing section.

How Some of the Factors Influence Multiple Sections of the Target Costing Process

Several of the factors influence more than one of the sections. These factors play a more sophisticated and pervasive role in shaping the target costing process. The intensity of competition is one of the most powerful factors. While it primarily influences the market-driven costing section, it also influences the other sections. In general, the more intense the competition faced, the greater the benefits derived from target costing as a whole. Therefore, for most firms, as the intensity of competition increases so will the energy expended on all three sections of the target costing process.

The number of products in the line, besides influencing the product-level target costing process, helps shape the market-driven costing process when product line rationing is necessary. Product line rationing occurs when customers demand a greater variety of products than the firm can afford to support. In this case, the market analysis the firm undertakes must identify the products the firm needs to launch if its overall profit objective is to be met. If the number of products has to be limited, the role of the target costing system shifts away from helping ensure individual product profitability toward helping identify the most profitable mix of products. Nissan uses computer simulations to ensure the right mix of products. Thus, target costing is especially beneficial for firms that have to limit the number of products they produce.

Product complexity, in addition to influencing the product-level target costing process, also shapes the component-level target costing process. As the number of components increases, so typically will the number of suppliers involved. Hence, as the complexity of the components increases, the need to monitor the suppliers increases. Therefore, firms that produce complex products typically have more developed component-level target costing processes.

The degree of innovation, in addition to influencing the product-level target costing system, also plays a role in determining the market-driven costing and component-level target costing processes. As the degree of innovation increases, it becomes more difficult to predict the location of survival zones as the customers' reactions to new products becomes more difficult to anticipate. For products with a high degree of innovation, the information required by the market-driven section is difficult to collect, while its value is high. Therefore, the amount of energy expended on this section is considerable. For example, Nissan and Toyota face this condition and have highly developed market-driven costing sections.

For products with low innovation, the information for the market-driven costing section will be easy to collect and of high value. Therefore, moderate energy will be expended on the section. This is the situation faced by Sony for most of its Walkman products, which are minor modifications of the products they replace. For such products, the selling price of the new product is primarily determined by the selling price of the product it replaces. A similar situation occurs for the component-level target costing section. As the degree of innovation increases, the ease of information collection falls. Therefore, the level of energy expended also increases because historical information which is easier to collect is no longer applicable.

Highly innovative products rely on new technologies and new components, forcing the firm to identify new suppliers on a regular basis. The application of new technologies, the acquisition of new components, and the reliance on new technologies all reduce the predictability of component costs.

The duration of product development not only increases the formality of the product-level target costing process, it also increases the importance of the other two sections. The market-driven section increases in importance because the risk is higher that customer requirements will change during the product development process. Therefore, continuous monitoring of customer attitudes and competitive offerings is necessary. Similarly, the longer development cycle means that suppliers have more time to experience changes such as input prices, and so additional component-level target costing is required to ensure that the suppliers remain on target.

COMPARING THE
TARGET COSTING PROCESSES

As the processes at the six companies demonstrate (Isuzu's target costing system was not documented), target costing is not a monolithic technique but an adaptive one. In addition to creating pressures to reduce costs, the target costing process creates a vital communications link between the marketing, engineering, purchasing, and manufacturing functions. Thus, target costing is more than simply a profit or cost management technique. It is a highly sophisticated way both to increase intraorganizational integration across a number of functions and to transmit the competitive pressure faced by the firm to its product designers and suppliers.

While all the systems at the six companies contain the three major sections—market-driven costing, product-level target costing, and component-level target costing—the process of target costing at the six firms is quite different. In particular, they expend different levels of energy on the three major srctions. To compare the overall target costing processes at the firms, each factor was assigned to an ordinal scale of plus or minus. If the relative effect of a factor was perceived as being favorable to target costing, the factor was assigned a plus value. If it was considered unfavorable, it was assigned a minus value. The result of this analysis was a plus and minus profile for each firm, shown in Table 9-1.

Each profile provides a way to observe the cumulative effect of the factors on each of the three major sections in the target costing process. If all the factors relating to a given section are favorable, then the firm is expected to have a well-developed process for that section of the target costing process.

The Target Costing Process at Toyota and Nissan

For Toyota and Nissan, all the factors favor target costing. The firms are in a confrontational industry with sophisticated consumers whose requirements are changing rapidly and who know what they want. In addition, the firms' product strategy and the characteristics of the product all suggest that the benefits of target costing will be high. Nissan and Toyota obtain a considerable part of their products from outside suppliers and subcontractors. Both

TABLE 9-1. THE RELATIVE VALUES OF THE INFLUENCING FACTORS

Factors	Nissan	Toyota	Komatsu	Olympus	Sony	Topcon
Intensity of competition	+	+	+	+	–	+
Nature of the customer						
Customer sophistication	+	+	+	+	+	+
Rate of change of requirements	+	+	–	+	–	+
Understanding of requirements	+	+	+	+	–	+
Product strategy						
Number of products	+	+	+	+	–	–
Frequency of redesign	+	+	+	+	+	+
Degree of innovation	+	+	–	+	–	+
Characteristics of the product						
Complexity	+	+	+	–	–	–
Up-front investments	+	+	+	–	–	–
Duration of product development	+	+	+	–	–	–
Supplier-base strategy						
Degree of horizontal integration	+	+	+	+	+	+
Power over suppliers	+	+	+	+	–	–
Nature of supplier relations	+	+	+	+	–	+

firms clearly dominate their suppliers. Suppliers are much smaller firms that typically rely heavily on Nissan and Toyota for their survival.

Both firms therefore put considerable effort into all three steps of the target costing process. They both have highly sophisticated procedures for monitoring their customers' changing requirements throughout the product development process. These procedures are integrated into the market-driven costing process. In addition, the product-level target costing process is quite elaborate, with the product design being modified continuously to ensure achieving the target cost. This modification is continued up until one month before launch when the final specifications of the standard version are established.

Finally, the component-level target costing process is highly sophisticated. The decomposition of the product into its major functions and the major functions into components reflects the ability of Nissan and Toyota to legislate purchase prices to their suppliers. In most cases, the target costs set for purchased parts become the suppliers' selling prices. Only under very specific circumstances is the target cost for a purchased part violated. However, both firms maintain highly cooperative relations with their suppliers and share considerable design expertise with them. While their component-level target costing systems apply intense cost pressure on their suppliers, the firms dedicate considerable resources to help their suppliers find ways to achieve their targets.

Therefore, it is not surprising that these two firms have mature, well-developed target costing procedures that are highly formal and extensive. The identified factors, however, do not capture all the forces that influence the process of target costing at the firms. Despite the fact that Toyota and Nissan are in the same industry and are direct competitors, their target costing processes are not identical. Other factors at the firm level also must play a role in determining the exact nature of the target costing process.

The Target Costing Process at Komatsu

For Komatsu, all but two of the factors favor target costing. The commercial nature of the product and the relatively slow changes in the requirements of customers mean that the degree of innovation in products is typically low. Komatsu often undertakes minor modifications to its products to keep them up to date as opposed to undertaking major redesigns. The low rate of change of customer requirements coupled with the extent to which customers understand their requirements means that sophisticated customer analysis is not required. The firm does not have to monitor customer requirements continuously or keep fine-tuning the design of their products. So Komatsu spends considerably less energy on the market-driven costing process than Toyota or Nissan, and the effect of this single factor on the market-driven costing process is significant.

The lower degree of innovation allows the firm to rely more heavily on the past than Nissan or Toyota. For example, the functional and productivity analyses that lie at the heart of the firm's target costing processes assume that history is predictive. Functional analysis assumes that the new model will be using the same technology as the old model, and productivity analysis assumes the same general production process will be used. Despite the relatively low degree of innovation, frequency of redesign is high, requiring a formal system that is well documented and communicated to the users. In addition, the high up-front investments and complexity of the products cause target costing to be of great benefit to Komatsu. Consequently, the product-level target costing process at Komatsu is as sophisticated as that at Toyota and Nissan but more internally focused.

The firm relies heavily on historical cost information to set the purchase prices of externally acquired components. Its relations with its suppliers are not as cooperative as those of Nissan and Toyota, and it relies more heavily on component-level target costing and less on other interorganizational cost-management techniques.

The Target Costing Process at Olympus

Olympus faces a confrontational environment with highly sophisticated customers whose requirements are changing rapidly. Reflecting the values of the factors that influence the market-driven costing process, that process is well developed and monitors customer preferences throughout the product development process. For Olympus, the nature of the product reduces the benefits of product-level target costing: cameras are much simpler than automobiles. In addition, they require lower up-front investments and have a shorter product development cycle. These differences cause the Olympus product-level target costing process to be simpler and less formal than that at either Toyota or Nissan. Instead of multiple decision points, the decision process at Olympus centers predominantly around the launch decision. The high degree of horizontal integration means that Olympus relies heavily on its suppliers, over which it exerts considerable power. Consequently, the component-level target costing process is well developed and as sophisticated as Toyota's or Nissan's.

The Target Costing Process at Sony

For Sony most of the factors are nonfavorable. The firm has developed a competitive advantage in the technology required to create ever smaller, long-playing Walkmans. This advantage has been sustained ever since the introduction of the Walkman. The firm is thus not fully in confrontation. In addition, the rate of change of customer expectations is relatively low, and the degree of innovation between each generation usually is small. Thus, only one of the factors that influence the market-driven costing process, customer sophistication, is favorable, and the firm expends little effort on market-driven costing. Typically, the selling price of the Walkman that is being replaced is used as the starting point for estimating the target selling price, and the target profit margin is assumed to be the same for both the old and new models.

The firm's product strategy results in relatively few models in the line and frequent redesigns but little innovation. Thus, only one of the product strategy-related factors favors product-level target costing. In addition, Walkmans are not particularly complex products, require low up-front investments, and have product development cycles measured in months, not years. Consequently, the product-level target costing process is simple and unsophisticated compared to the other firms in the sample.

The firm provided little information about its supplier-base strategy. However, it appears that the strategy is not particularly favorable toward component-level target costing. With all these factors not favoring target costing, it is not surprising that Sony has the least well-developed target costing process of all the firms. It is important to realize that it is not the relative value of a single factor that leads Sony to adopt a simple approach to target costing; rather it is the cumulative effect of many of the factors being relatively unfavorable.

The Target Costing Process at Topcon

For Topcon several factors are considered relatively unfavorable to target costing. The factors that influence the market-driven costing process are all favorable, and the firm does expend considerable energy on developing allowable costs. Of the six factors that influence the product-level target costing process, four are relatively unfavorable. Consequently, the product-level target costing process is relatively unsophisticated.

Topcon produces only a small number of ophthalmic instruments a year compared to the number of units that the other firms sell. This small volume is insufficient for Topcon to wield any significant power over its suppliers. Therefore, developing target costs for components is not particularly valuable for Topcon. Since it cannot control supplier costs by setting target costs for purchased parts, it has to live with the prices set by its suppliers. So the focus of the value analysis program is to find ways to reduce costs by changing the design of the products and, when possible, by negotiating with suppliers, not by setting their prices.

The remaining aspects of the target costing process at Topcon are less well developed, reflecting the lower value that target costing has for that firm. For example, the discipline of target costing is not as intensely applied at Topcon as it is at the firms already discussed. In addition, the target costing process is part of the firm's turn-out-value (TOV) system and not a stand-alone system as at the other firms. The TOV is also responsible for product costing and other cost-management techniques.

The Plus/Minus Profiles

The plus/minus profiles help explain why the target costing processes at the six firms differ. Whenever one or more of the factors are considered relatively unfavorable, the amount of expended energy and the sophistication of the process decrease. It is the cumulative effect of all the factors that shapes the target costing process at the firms, not just a single factor. Each of the factors identified primarily influences one of the three major sections in the target costing process. These three processes can be influenced independently of each other, so it is possible for some of the sections to be sophisticated while others are relatively simple.

SUMMARY

By comparing the intensity of competition, the nature of the customer, the firm's product strategy, the characteristics of the product, and the supplier-base strategy it is possible to explain the nature of the target costing process. These factors influence the target costing process because they help determine the benefits that the firm will derive from applying target costing and hence the energy the firm is willing to expend on each part of the process. Presumably if the benefits fall below a certain level, then the firm will either never implement a target costing system or discontinue its use.

The intensity of competition influences the target costing process, especially when the basis of competition is enhanced product functionality. All the firms studied are in intensely competitive industries, and most, if not all, have adopted confrontational strategies. The only possible exception is Sony, which is on the edge of being a differentiator due to its sustainable technological lead over its competitors.

The nature of the customer covers the degree of sophistication of the customer, the rate at which customer requirements change, and the degree to which customers understand their future requirements. These three factors shape the target costing process because they help determine the benefits derived from market-driven costing. Customer sophistication determines the width of the survival zones of new products. The narrower the zones, the more beneficial target costing is. The rate of change of customer requirements captures how fast survival zones are moving over time. The faster the zones are moving, the more beneficial target costing becomes as it increases the probability that new product designs will reflect customer preferences when launched. The customers' degree of understanding of their future requirements also influences the benefits derived from market-driven costing. As the customers' ability to estimate future requirements drops, it becomes more difficult to provide the product designers with adequate information on the type of products they should be designing.

A firm's product strategy was identified as an influencing factor because, among other things, it decides the number of products in the line, the frequency of redesign, and the degree of innovation. These factors are important because they help determine the benefits of product-level target costing. The number of products affects the benefits because target costing operates at the product level, and the more numerous the products, the greater the benefits are likely to be. Similarly, the frequency of redesign is important because it determines how often new products are introduced. The more frequently products are introduced, the greater the benefits of product-level target costing. The degree of innovation has to be considered because, to a certain extent, target costing relies on historical cost information. As the degree of innovation increases, the value of historical information drops. Therefore, evidence suggests that the benefits of product-level target costing fall with increases in the degree of innovation.

The characteristics of the product are important because they help determine the complexity of the product, the magnitude of up-front investments, and the duration of the product development process. These factors play a role in determining the benefits of product-level target costing because they shape the product design process. The complexity of products helps determine how formal the target costing process should be. With simple products an informal process is probably adequate. However, complex products such as automobiles and bulldozers require more formal target costing processes.

The magnitude of the up-front investments is critical because it determines whether the firm can afford to launch numerous products with a high probability of failure or a limited number of products that must have as high a probability of success as possible. The higher the up-front investment, the greater the benefits of product-level target costing. Finally, the duration of the product development process plays a role because it helps determine the risk that the product will be outside its survival zone when launched. The longer the product development process, the greater the risk, therefore, the greater the benefits of target costing and the more formal the target costing process.

One of the major outputs of target costing is component-level target costs. These are the prices the firm is willing to pay for externally acquired components. Such target costs appear to have maximum value when the firm outsources a high proportion of the total cost of its products externally, when it has sufficient power over its major suppliers to dictate selling prices, and when supplier relations are cooperative as opposed to adversarial.

Several of the factors influence more than one of the sections of the target costing process. In particular, the intensity of competition influences all three sections because as profits decrease, the need to launch successful products increases. When the number of products in the line is limited, market-driven target costing becomes more important because identifying the right product mix becomes even more critical. In addition, product complexity influences the component-level target costing process because the more complex the product, the greater the role of suppliers in achieving the target cost. Finally, the duration of the product development cycle influences both the market-driven costing and component-level costing processes because as the time to develop a product increases, so does the risk that conditions either in the market or at suppliers will change.

All the above factors help determine the magnitude of the benefits a firm can derive from target costing. When the benefits are low, the target costing process is usually less sophisticated, and less energy is expended on the process. At Toyota and Nissan the target costing systems are formal and complete because virtually every factor supports target costing. In the other firms at least some of the factors do not favor target costing. The existence of these factors helps explain the concentrated decision process at Olympus, the internal focus of the target costing system at Komatsu, and the limited systems at Topcon and Sony.

CHAPTER 10

TARGET COSTING AND
VALUE ENGINEERING IN ACTION

INTRODUCTION

We will use a hypothetical company, the Acme Pencil Company Ltd., to illustrate target costing and value engineering in action. Acme wants to introduce a new graphite pencil that differs from existing products because of its enhanced functionality. Despite the simplicity of the product, the example captures much of the richness of the target costing process from market analysis to supplier negotiations.

THE ACME PENCIL COMPANY LTD.

Acme produces a wide variety of writing instruments including traditional graphite pencils, ball-point pens, highlighters, and fountain pens. It is a well-established firm with a reputation for innovation. Acme has four major competitors, all essentially evenly matched. None of these firms can point to a sustainable competitive advantage.

Their products are very similar and they compete for the same customers, both commercial and retail.

Feedback from customers indicates a functionality problem with the firm's traditional graphite pencil-with-eraser products. The erasers on the existing designs usually run out about two-thirds of the way through the life of the pencil. The user then has to find a separate eraser, live without the ability to erase, or throw out the remaining one-third of the pencil. The average consumer is aware of the problem, though most of them are not overly concerned. However, interactions with customers suggest that a long-life eraser guaranteed to last as long as the pencil could be used to create a new high-end product. This enhancement is attractive because it increases by 50% the effective life of the pencil (3/3 is 50% higher than 2/3).

These interactions with customers identify the relative importance of features that the customers value. The most important feature is the ability of the pencil to write well. This ability depends only on the lead, and all graphite pencils on the market have equivalent performance. The second most important feature is the ability to erase. The new, long-life product will have a much higher performance in this respect. The final feature is appearance. The low-end products have a less attractive appearance than the high-end products. The only difference between the existing high-end product and the new long-life one is the eraser performance. The feature weightings of the two products, based on a scale of 10, are shown in Table 10-1. Since the competitors' products are identical to the existing high-end ones, there is no need to undertake feature analysis on their products.

TABLE 10-1. FEATURE WEIGHTING FOR LONG-LIFE
            AND EXISTING HIGH-END PRODUCTS

	Product	
Feature	Long-life score	High-end score
Writes well	10	10
Erases well	7	5
Looks attractive	5	5

The product engineers have reacted enthusiastically to the concept of the new long-life eraser product and have begun to experiment to see if they can find a simple way to overcome the limitations of existing designs, for example, by using a physically longer eraser. So far, they have failed to find an easy solution. The longer eraser has a tendency to break almost immediately. So the engineers are beginning to look at more fundamental ways to extend the life of the eraser, including the use of new polymers. Their objective is to create a pencil that can write 10,000 ft. of drawn lines and erase 3,000 ft. A 30% eraser-to-write capability is considered appropriate for the new long-life product as that is the average use ratio. The existing products can write for 10,000 ft. but can erase only 2,000 ft.

The new product will be part of the firm's traditional graphite pencil product line. This line represents 32% of the firm's sales and currently contains three sub-lines. Each sub-line consists of a full range of pencils with leads that have different degrees of thickness and hardness. The different variants in each sub-line are sold at the same price and provide horizontal differentiation of the product within the sub-line, each level of hardness and thickness serving a different purpose. The three sub-lines are vertically differentiated and provide different levels of quality and functionality at different prices. The cheapest sub-line has no eraser and uses a single-color mat paint. The middle sub-line has an eraser and uses the same single-color mat paint, though different colors are offered. The most expensive sub-line has an eraser and uses multicolor glossy paint. The new long-life product is expected to create a fourth sub-line that will sell at an even higher price than the current high-end one.

Acme has had a well-developed target costing system in place for just over five years. All new products are subjected to the discipline of target costing. If a product cannot achieve its target cost, it is not introduced. The new long-life eraser pencil is no exception. If there is no way to design the product so it can be manufactured at its target cost, it will not be introduced.

Acme's target costing process can be broken into three major sections. The first part is market-driven costing, in which the allowable cost of the product is established. The second section, product-level target costing, establishes the target cost of the product. The gap between the allowable cost and the target cost highlights the strategic cost-reduction challenge. In the final part, component-level target costing is used to set the target costs for each component in the product.

MARKET-DRIVEN COSTING

The Process

The firm's long-term sales and profit objectives are based on five-year projections of what top management thinks is realistically achievable if everyone in the firm expends significant efforts to reach their goals. The profit objective for the next five years for the graphite pencil line is 12%, and the sales volume objective is 2 million dozen units. To help achieve this objective, the new product line is expected to earn above-average profits and have sales levels between 100 and 200 thousand dozen units in the first year, with subsequent years' sales being at least 200 thousand dozen units.

The retail list selling prices of the existing products are $2.45, $3.00, and $3.50 per dozen for the low-end, medium, and high-end products respectively. The actual selling price varies depending on the nature of the store selling the product. Small stationers typically sell products at prices close to list, while the big discount chains sell at about 45% of list. Acme sells directly to the large chains and via distributors to the smaller firms. The ratio of retail list price to Acme's average selling price is roughly constant for all products in the graphite pencil product line. The average wholesale selling prices across both channels for the three graphite pencil sub-lines are $0.70, $0.85, and $1.00 per dozen, respectively, providing profit margins of 10%, 11.3%, and 13% (see Table 10-2). These margins have been relatively stable over the years.

The new product is expected to have a retail list price of between $4.02 and $5.25. The lower price limit is set to reduce excess cannibalization of the existing high-end products. Long experience in the pencil industry suggests that customers are indifferent to price differences below 15%, making it impossible to vertically differentiate products effectively with price points closer than 15%. If Acme prices the new product below $4.02 ($3.50, the retail list price of the high-end line, times 115%, the minimum sustainable price difference), it would have to take the existing high-end sub-line off the market. Top management is reluctant to take such a risky step at this time because the high-end products are the most profitable of the three existing sub-lines and are considered key to the graphite pencil product line's ability to reach its long-term profit objectives.

The problem with the lower price limit is that it is unlikely to generate adequate profits for the firm, even if the sales volumes are higher than expected. Even though the new design is still incomplete, the product engineers have indicated that a significant increase in unit product costs is likely for the new product. Since the product will require significant up-front investment, a profit margin well in excess of 12% is required for the firm to achieve its profit objectives. At the lower price limit, the ability of the firm to achieve such a high margin is considered highly questionable.

The upper price limit of $5.25 reflects the increased value that customers are likely to perceive in the new product. Since many customers will now be able to use the last third of the pencil, a 50% increase in price is considered appropriate for setting the upper price limit that customers are willing to pay. The cost of using a separate eraser is comparable to the price increase but in addition requires

TABLE 10-2. THE SELLING PRICES, PROFIT MARGINS, AND
        MANUFACTURING COSTS PER DOZEN
        OF THE EXISTING PRODUCTS

	Product		
	Low-end	Medium	High-end
Retail list selling price	$2.45	$3.00	$3.50
Average wholesale selling price	$0.70	$0.85	1.00
Profit margin	10%	11.3%	13%
Total cost	$0.67	$0.75	$0.87

the extra effort of finding one. Therefore, the value to the customer of avoiding that extra hassle is higher than the upper price limit. For customers who don't care about throwing away the last one-third of the pencil, the new long-life product has no extra value, so they are unlikely to switch from the pencils they currently use to the new ones.

The problem with the upper price limit is that sales volumes at that price are expected to be too low for the firm to achieve its profit objectives even given the high margins. Market analysis indicates that only 10% of potential customers are expected to buy the new product at the upper price limit. The majority of potential customers have indicated that at the upper price limit, they prefer to buy low-end mechanical pencils rather than conventional ones, even with enhanced eraser functionality.

After considerable analysis and discussion, the target retail list price is set at $4.25. This target selling price is selected for three reasons. First, top management believes that customers will feel that they are getting good value for their money. The perceived value for money is important because top management views the products as having a significant role in reinforcing the firm's reputation as an innovator. While no firm can claim to be the technological leader, Acme believes that it has a slight technological advantage over its competitors. Being the first to market with innovative products plays a critical role in giving the firm a small competitive advantage over its competitors.

Top management also believes the target selling price is low enough to ensure acceptable market penetration because they expect a large percentage of customers to shift from mechanical pencils to the new product. Low-end mechanical pencils are not refillable and have erasers that also usually wear out before the pencil runs out of lead. Their retail list price starts at $5.00 per dozen. Customer analysis has indicated that the $0.75 ($5.00 – $4.25) price break is sufficient to cause about 30% of low-end mechanical pencil users to shift to the new high-end graphite pencils. Converting customers from mechanical pencils is particularly attractive for Acme since it doesn't manufacture those products.

Finally, management believes that the target selling price is high enough to prevent cannibalization of the existing high-end product. They expect cannibalization to be low because only about 15% of current high-end traditional pencil users are expected to switch to the new product. If they were all current customers of the firm, the result would be a significant loss of profitability for the existing high-end pencils. However, since at least half these users are expected to come from competitors' customers, the predicted loss of sales volume for the existing high-end sub-line is considered acceptable.

If Acme achieves a 30% conversion rate from mechanical pencils and a 15% conversion rate from existing high-end traditional pencils, the target sales volume of the new product is expected to be 150 thousand dozen units in the first year, rising to 250 thousand dozen in subsequent years and giving total sales of 1,150 million dozen over the five years. These estimates are considered realistic and capable of supporting the firm's long-term sales objectives. Critically, since the conversions are expected to come about evenly from each competitor, their individual market share losses will be quite modest, and they are unlikely to react immediately to the introduction of the new product. Top management expects me-too products to appear toward the end of the second year, causing the retail list selling price to drop to $4.05 per dozen. The $4.25 target retail list selling price is thus expected to be sustainable for the first two years of the product's five-year life.

If the firm sells the new product at the same ratio of retail list price to wholesale price as the existing high-end sub-line, then Acme's target wholesale selling price of the long-life pencil will be approximately $1.22 ($4.25, the target retail list price of the long-life pencil divided by $3.50, the retail list selling price of the high-end product, times $1.00, its average wholesale selling price). Once the firm sets the target wholesale selling price and estimates a target volume, they can establish the target profit margin. The firm's starting point is the 12% average set for the traditional graphite pencil product line. Top management is pressing for a higher target profit margin than the average for all pencils.

The target profit margin calculation begins with a projection of historical profit margins against price. Historically, as the retail list price of a pencil increases, so does the profit margin it generates. This relationship is curvilinear, not linear, and extending the curve to $1.22, the target retail list selling price, indicates that the new pencil should earn a profit margin of 15% (see Figure 10-1).

The up-front investment of $75,000 ($50,000 for research and development and $25,000 for launch) represents a cost of $0.065 per dozen ($75,000, the up-front investment, divided by 1,150,000 dozen, the total sales volume anticipated over the life of the product) over the expected life of the product at its target sales volume. To recover the up-front investment thus requires an additional profit margin of 5.3% ($0.065, the up-front investment cost per dozen, divided by $1.22, the target wholesale selling price).

However, the 15% profit margin calculated using the historical trend between selling price and profit margin already includes an allowance for up-front investment. For the existing product lines, the up-front investments run at about 50% of that anticipated for the new product. Thus the incremental profit margin required to

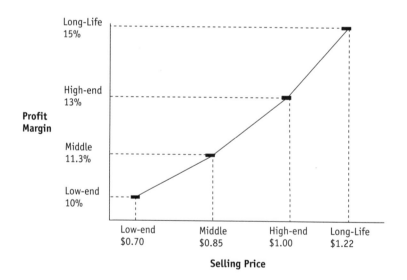

FIGURE 10-1. ACME PENCIL: PROFIT MARGIN VS. SELLING PRICE

cover the up-front investment for the long-life pencil is 2.65% (50% of 5.3%, the additional profit margin required to cover the up-front investment). If the product is to earn its 15% margin and recover the additional up-front investment, it must generate a margin of 17.65% (15% plus 2.65%). Unfortunately, this margin is considered unachievable.

After considerable debate, the target profit margin is set at 16%, which represents a compromise between the 15% predicted from the existing products' profit performance and the firm's long-term profit objectives. Simulations of the expected profits for graphite pencils show that at 16% the average profit margin for the first year will be just under 12% (see Figure 10-2). However, kaizen cost reductions over the life of the long-life product are expected to create a higher margin in the last few years. These higher margins are considered adequate to offset the up-front investments not captured in the 16% and increase the average margin of traditional

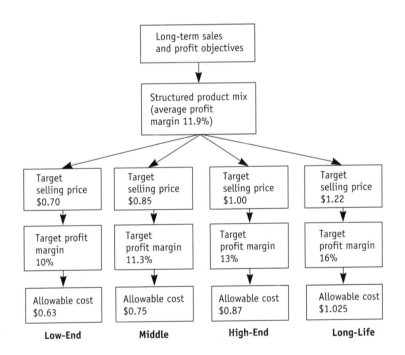

FIGURE 10-2. ACME PENCIL: MARKET-DRIVEN COSTING

pencils to over 12% for the last three years of the product's life. All concerned in the target costing process agree that, while aggressive, the target profit margin of 16% is achievable.

A 16% profit margin for the long-life pencil means that the allowable cost to sales ratio is 84% (100% minus 16%) of the firm's selling price. This profit margin thus results in an allowable cost of $1.025 ($1.22, the target average wholesale selling price, times 0.84, the cost to sales ratio). This is the cost at which the new product must be manufactured if it is to earn its target profit margin at its target selling price.

An Analysis of the Process

At Acme, the role of the customer in the price-driven costing process is two-fold. First, customers were instrumental in identifying the improved functionality at the heart of the new product. When asked, they were able to identify the improvements they wanted in the functionality of Acme's products. Given the nature of the product, there is little opportunity to introduce revolutionary functionality that the customer doesn't expect. So at Acme the chief engineer plays a relatively small role in setting the themes of the new product.

The second role customers play is in setting the range of feasible selling prices. The upper price limit is determined in part by the incremental value that customers perceive for the new functionality. The lower price limit is set by customers' ability to differentiate between the selling prices for the long-life and existing high-end products.

The role of competitors in the price-driven costing process is defined by the way they influence customer expectations. Acme is not expecting its competitors to launch new products that shift the survival zones of existing products, so the primary role competitors play is in the way their competitive products help define the target selling price. The selling price of mechanical pencils influences the upper price limit because the price differential between the long-life pencil and the lowest-priced mechanical pencils must be sufficient for the customer to be willing to switch to a traditional pencil. The competitors' products that are equivalent to the firm's existing high-end products help determine the lower price limit because even if Acme withdrew their high-end products, the competitor's equivalent products would set the lower price limit.

The target selling price of $4.25 is set to make a trade-off between the market share Acme wants to achieve, the degree to which they want to strengthen their corporate image as an innovator, and the long-term profitability of the product. This trade-off is a complex one that can be made only in light of Acme's strategy and profit objectives. The market share objectives have to allow for cannibalization. There is no point in setting an aggressive market share objective for the new long-life pencil if the market is predominately from existing high-end customers. Here, customer loyalty may work against the firm. However, since customer loyalty in the industry is known to be low, the bulk of sales for long-life pencils is expected to come from new, not existing, customers. The trade-off is further complicated by the need to take into account the reactions of competitors. If the long-life pencil turns out to be very successful, competitors will be forced to react and may cause the overall profitability of the product to fall by introducing me-too versions or dropping prices of existing products.

The target profit margin is based on a number of factors. Acme uses the historical relationship between selling price and profit margin to establish the starting point for setting the target profit margin for their new product. The next step is to incorporate life-cycle costs into the target profit margin. The final target profit margin of 16% incorporates the up-front investment, changes in the selling price over time, and changes in the manufacturing cost over time. In addition, it reflects the firm's ability to earn the target profit margin. Such realism is critical if the firm is to achieve its long-term profit objectives because setting unrealistic target profit margins would jeopardize the firm's ability to achieve its long-term profit objectives.

Acme's adjustment of its target profit margin to achievable levels causes the allowable cost to be different from the benchmark cost. If the most efficient producer in the industry can achieve the

17.65% margin, then the benchmark cost would be $1.00 ($1.22, the target selling price, times (100 – 17.65)/100, the cost to sales ratio). Thus, the firm has already accepted a cost disadvantage of $0.025 ($1.025, the product's allowable cost minus $1.00, the product's benchmark cost). This disadvantage is essentially hidden by selecting 16% as the target profit margin.

PRODUCT-LEVEL TARGET COSTING

The Process

At Acme, the product-level target cost for the new long-life pencil is set in an iterative process in which the value engineering team integrates supplier cost quotes with the cost estimates provided by manufacturing. The process begins by determining the current cost of the new pencil—the cost at which it could be produced by buying the components at today's costs and using existing production processes.

The pencil has five components: the graphite "lead," the wooden body, the paint, the eraser, and the metal band that holds the eraser to the body of the pencil. The new long-life pencil requires a heavier band and a more expensive eraser. The other components are virtually identical to those used in the existing high-end product. Using available cost data and best estimates from the suppliers of the new metal band and eraser, the current cost of the components is estimated to be $0.60 per dozen compared to $0.35 per dozen for the existing high-end product (a complete breakdown of component costs for both products is shown in Table 10-3). The component cost for the new product is particularly high because the extra-heavy metal band required by the new longer-life eraser adds $0.15 per dozen, and the new eraser adds another $0.10 per dozen to the cost of the existing high-end pencil.

The manufacturing process has seven major steps.

- The graphite "lead" is produced by extruding a graphite and clay slurry to form the lead.
- The wood portion is manufactured by planing cedar planks into "half-pencils" many times the length of a pencil.

TABLE 10-3. CURRENT COST OF COMPONENTS FOR LONG-LIFE
AND EXISTING HIGH-END PRODUCTS

	Product	
Component	Long-Life	High-End
Paint	$0.05	$0.05
Body (wood)	0.05	0.05
Lead (graphite)	0.06	0.06
Band (metal)	0.20	0.05
Eraser	0.24	0.14
Total	$0.60	$0.35

- A semicircular groove for the graphite lead is cut into the wood.
- The wood halves and the lead are assembled and glued together.
- The resulting "pencil" is painted.
- The pencil is cut to size.
- The eraser is attached through a crimping process using the metal band.

The manufacturing cost used to compute the target cost is not the same as the manufacturing cost reported by the firm's cost system. The cost system uses accelerated depreciation charges to reflect the firm's tax policies. The product designers consider the resulting depreciation charges for the special equipment required by the new product excessive for target costing purposes. Consequently, they recalculate the depreciation charge for all major equipment in the production process using a straight line approach and a longer, more realistic estimate of the equipment's economic life. Given the age of the equipment, the result of this correction is a 5% reduction in reported manufacturing cost. The current cost of manufacturing is set at $0.625 per dozen (a complete breakdown of manufacturing costs is shown in Table 10-4). Summing the component and manufacturing costs identifies a current cost of $1.225 per dozen (see Table 10-5). The cost-reduction objective obtained by subtracting the allowable cost from the current cost is thus $0.20 per dozen ($1.225, the current

TABLE 1 0-4. CURRENT MANUFACTURING COSTS OF LONG-LIFE
AND EXISTING HIGH-ENDª PRODUCTS

	Product	
	Long-Life	High-End
Manufacturing Steps		
Extrude graphite lead	$0.10	$0.10
Plane wood	$0.125	$0.15
Cut groove	$0.05	$0.05
Assemble and glue	$0.10	$0.10
Paint	$0.10	$0.10
Cut to length	$0.02	$0.02
Attach eraser	$0.18	$0.05
Total	**$0.625**	**$0.52**

cost per dozen minus $1.025, the allowable cost per dozen). This objective is shown in Figure 10-3.

The as-if cost is set at $1.175 per dozen. This is the cost at which the long-life pencil could be manufactured today if no additional cost-reduction savings are introduced beyond those already recognized but not yet incorporated into existing products (see Table 10-6). The as-if cost incorporates two as yet to be implemented cost-reduction steps. First, the firm has recently identified a new way to paint pencils that will be $0.03 per dozen cheaper. Of this $0.03, $0.01 comes from reduced paint costs and $0.02 from reduced painting costs. These savings are achieved by recycling the unused paint and spraying it again. A second savings of $0.02 per dozen is due to anticipated price reductions from the firm's suppliers. The suppliers

TABLE 1 0-5. CURRENT COSTS OF LONG-LIFE AND
EXISTING HIGH-END PRODUCTS

	Product	
	Long-Life	High-End
Cost Element		
Component costs	$0.60	$0.35
Manufacturing costs	$0.625	$0.52
Current Costs	**$1.225**	**$0.87**

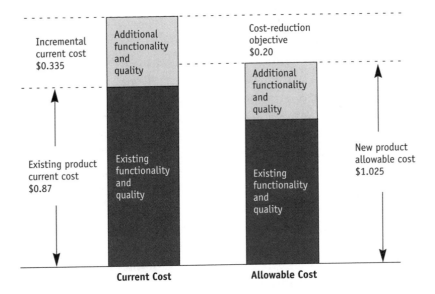

FIGURE 10-3. THE TARGET COST-REDUCTION OBJECTIVE

TABLE 10-6. CURRENT AND AS-IF COST
OF NEW LONG-LIFE PRODUCT

Components and Manufacturing Steps	Products		
	Long-Life	Long-Life	Long-Life
	Current Cost	As-If Cost	Difference
Components			
Paint	$0.05	$0.04	$0.01
Wood body	$0.05	$0.04	$0.01
Graphite lead	$0.06	$0.05	$0.01
Metal band	$0.20	$0.20	$0.00
Eraser	$0.24	$0.24	$0.00
Subtotal components	$0.60	$0.57	$0.03
Manufacturing Steps			
Extrude graphite lead	$0.05	$0.05	$0.00
Plane wood	$0.125	$0.125	$0.00
Cut groove	$0.05	$0.05	$0.00
Assemble and glue	$0.10	$0.10	$0.00
Paint	$0.10	$0.08	$0.02
Cut to length	$0.02	$0.02	$0.00
Attach eraser	$0.18	$0.18	$0.00
Subtotal manufacturing	$0.625	$0.605	$0.02
Total Cost	**$1.225**	**$1.175**	**$0.05**

for the lead and wood for the body have indicated that they will be able to reduce their prices by $0.01 each.

Normally, the firm's kaizen costing system also would have reduced the cost of production. However, there is general concern that introducing the new product will disrupt manufacturing sufficiently that it is not wise to anticipate any manufacturing cost savings for the long-life product in the first year. In subsequent years, the cost of production is expected to fall by about 3% per year.

The design team thus faces a cost-reduction objective of $0.15 per dozen ($1.175, the as-if cost per dozen, minus $1.025, the allowable cost per dozen). The most likely place to find these savings, according to the product designers, is by redesigning the eraser and metal band. These two components, at $0.24 and $0.20 respectively per dozen, represent some 73% of total component cost.

The design team starts with the metal band and explores the possibility of using plastic instead of metal. Plastic costs less, but unfortunately it cannot be crimped like metal so it is more difficult to attach to the pencil body and eraser. The value engineering team finds a recently developed type of plastic that can secure the eraser to the pencil body for the life of the pencil. Using this plastic reduces per dozen costs by $0.04, after allowing for an additional $0.01 per dozen increase in attaching costs. The new plastic band also allows the eraser costs to be reduced by $0.08 per dozen. The plastic band can be molded into more complex shapes than the old metal one. This flexibility allows the diameter of the eraser to be smaller than the pencil's, thus reducing the amount of eraser required (see Figure 10-4). When the new design is tested with a group of customers, they are satisfied with the new design. The slimmer eraser works as well as the old design, and some of the customers even remark that they prefer it because it can be applied more precisely.

The combined savings of $0.12 per dozen ($0.04 from the metal band and $0.08 from the eraser) brings the drifting cost to $1.055 ($1.175, the as-if cost per dozen, minus the identified savings of $0.12 per dozen). This cost is within $0.03 of the allowable cost. Negotiation with the supplier of the glossy paint identifies an addi-

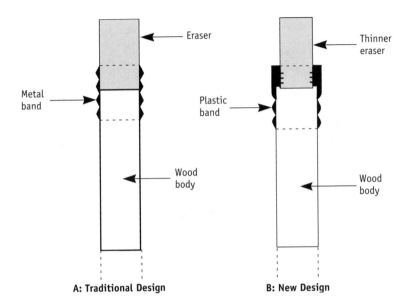

FIGURE 10-4. REDUCING THE COST OF THE NEW ERASER

tional $0.01 per dozen savings if the same type of paint is used for both the existing high-end and the new long-life pencils. No further savings can be found, so the product-level target cost is set at $1.045. This product-level target cost establishes the strategic cost-reduction challenge at $0.02 ($1.045, the target cost per dozen, minus $1.025, the allowable cost per dozen) and the target cost-reduction objective at $0.18 ($1.225, the current cost per dozen, minus $1.045, the target cost per dozen). These relationships are shown in Figure 10-5.

Despite the size of the strategic cost-reduction challenge, which represents a reduction in profit margin of 1.6% ($0.02, the strategic cost-reduction challenge per dozen divided by $1.22, the target wholesale selling price), no attempt is made to revisit the target pricing or functionality decisions. The original target price analysis is still considered valid, and the simplicity of the product makes it difficult to reduce functionality (the life of the eraser) without making the new pencil functionally identical to the existing high-end product.

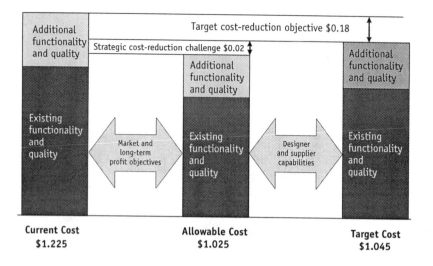

FIGURE 10-5. ACME PENCIL: THE STRATEGIC
COST-REDUCTION CHALLENGE

Setting the product-level target cost at $1.045 means that the target profit margin will shrink from the planned 16% to 14.4% (16%, the target profit margin, minus 1.6%, the forgone margin). This reduction in profit margin is considered acceptable for several reasons.

- The new product is considered strategic. If successful, it will enhance the image of the firm and perhaps cause the sales volume of other products to increase.

- Kaizen costing will increase profit margins while the new pencil is in mass production.

- The 14.4% is well above 12% and sufficiently close to the 15% predicted from the existing sub-line profit margins that many believe it to be more realistic in the first year than 16%. Had the resulting margin been below 14%, the product probably would have been withdrawn at this stage of its development.

An Analysis of the Process

The product-level target costing process introduces the cost-reduction capabilities of the firm's designers and suppliers into the target costing process. The starting point of the product-level target costing process is the current cost of the long-life pencil. This cost is derived from available information on the cost of the existing high-end product and the estimated costs of the new components and production processes to be used in the long-life pencil. The current cost is immediately adjusted for savings that have already been identified but not yet implemented. The resulting cost is the as-if cost. The next step is value engineering, which analyzes the product design to see if its cost can be reduced further through redesign. Acme value engineers substituted plastic for metal, found ways to reduce material content, and increased parts commonality, all of which are common value engineering practices.

The gap between the current cost and the allowable cost determines the overall cost-reduction objective. As the value engineering process continues, this objective is split into two parts, the target cost-reduction objective and the strategic cost-reduction challenge. The target cost-reduction objective is the portion of the overall cost-reduction challenge that the firm's engineers view as achievable, and the strategic cost-reduction challenge identifies the portion they view as unachievable. The engineers must negotiate with top management before the inability to achieve the allowable cost is authorized and the product-level target cost is established. Once established, this target cost is subject to the cardinal rule of target costing. The process of setting the strategic cost-reduction challenge must be disciplined and respected enough so that everyone views the identification of the challenge as a way to set achievable product-level target costs that will maintain the pressure on the product designers and suppliers, not as a way to create slack to make people's lives easier.

Setting a target cost higher than the allowable cost ensures that the product designers have a high probability of achieving the target cost. A high probability of achievement is necessary if the discipline of target costing is to be maintained through the cardinal rule. Inability to achieve the target cost must be perceived as a major failure of the design process, not simply an outcome of setting impossible targets. The strategic cost-reduction challenge identifies how far short of its long-term profit objectives the product will fall. Typically at this point in the process, the target price and functionality of the product would be reexamined. At Acme, such a reevaluation is considered unnecessary. Market conditions have not changed significantly during the process, and reducing the functionality of the product is considered inappropriate. Top management has to decide if the lower margin is acceptable. If it is not acceptable, then the product will not be introduced. Since they decide that the challenge is not too large, the product is approved for the next stage of the target costing process.

COMPONENT-LEVEL TARGET COSTING

The Process

Once the product-level target cost is established, it can be decomposed into component-level target costs. The interactions with the paint supplier, plastic band extruder, and eraser manufacturer have already established their selling prices, which are the target costs of these components. The cost of the paint ($0.03) was negotiated when the design team explored the idea of using the same paint as for the existing high-end pencils. The plastic band manufacturer has been rewarded for the creativity demonstrated in designing the new band with 100% of the volume for that component at a negotiated price of $0.15 for the first two years of the product's life. At the end of the two years, the part will be released for open bid from all the certified suppliers that produce equivalent products. The new eraser material is available from only one source, its inventor, and there is no room to negotiate its price ($0.16).

The target costs of the other two components, the graphite and the wood, are still to be negotiated, and the usual annual reduction had been anticipated when establishing the product-level target cost. However, the graphite suppliers all refuse to sell at any price lower than last year's. Their costs of material and energy have increased and cannot be offset by additional cost reductions on their part. The price increases have been so severe that simply maintaining the current price ($0.06) represents a significant achievement on their part. Accepting this price increases the drifting cost of the product by $0.01 to $1.055, since the anticipated savings will now not be realized.

Under the discipline of the cardinal rule, additional cost savings of $0.01 have to be found elsewhere to bring the drifting cost back to the target cost. The only remaining opportunity is the wood for the body of the pencil, so negotiations with the wood suppliers intensify. A value engineer from Acme visits the plant of one supplier and identifies ways to reduce their production costs by the necessary amount. With the negotiations completed and the target cost for the body established ($0.03), the drifting cost is now equal to the target cost ($1.045).

During the entire target costing process, the design engineers' progress is being monitored by graphing the drifting costs as they achieve each cost-reduction improvement (see Figure 10-6). The completion of supplier negotiations marks the shift to mass production. The hand-built prototypes have indicated that manufacturing costs will be identical to those of the existing high-end product. The new plastic band takes a little longer to attach, but the extra costs of this process are expected to be offset by kaizen savings in other processes.

An Analysis of the Process

The completion of the product-level target costing process signals the shift to component-level target costing. However, note that the firm's suppliers have already been involved earlier in the target costing process. Their estimates and bids have already been used to help set the product-level target cost. The cardinal rule plays a critical role during the component-level target costing process to help ensure achievement of the target cost. As the various suppliers indicate

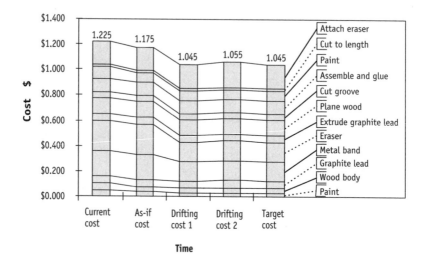

the prices at which they will supply their components, the drifting cost is recomputed.

The identification of the supplier of the polymer for the new eraser reflects the firm's strategy of extending its supplier base when it encounters innovative new suppliers. The firm's supplier-incentive scheme rewards the plastic band manufacturer for the new design and induces creativity in all the firm's suppliers. In return for the sole-source contract, the supplier shares the innovation with Acme. This design innovation will be shared with all the firm's plastic component suppliers so that they can incorporate the savings into their bids for the part when it is next up for bid. The awarding of the wood contract to two suppliers instead of only one is intended to maintain the firm's existing supplier base.

The other suppliers are expected to accept lower selling prices based on their historical cost-reduction rates. However, due to input price changes, they refuse to accept the business at the component-level target cost. Since all the firm's suppliers of graphite refuse, the target cost of that component has to be increased to realistic levels. As the target cost of one component is increased, the target cost of at least one other has to drop to maintain the product-level target cost. This search for savings elsewhere in the product illustrates the cardinal rule in action.

The value engineer's visit to the wood suppliers is the kind of action that value engineers at many of the firms take. It demonstrates the importance of maintaining cooperative supplier relations. Acme's objective is to attain the target cost, not destroy its supplier base. Helping a supplier find a way to sell a component at its target cost and still make an adequate return is inherent in, not counter to, the spirit of target costing.

Component-level target costing ends when the suppliers sign binding contracts for the components they supply. These contracts fix the purchase prices of the externally supplied components. Since Acme acquires all components externally, it can now compare actual component costs to target costs. The achievement of the manufacturing portion of the product-level target cost will become apparent only after about three months into mass production when the manufacturing process becomes stable.

SUMMARY

While a pencil is a very simple product, the application of target costing at the fictitious Acme Company illustrates the major issues that arise during the target costing process. The process begins by using market-driven costing to set the allowable cost of the new product. It continues by applying product-level target costing to create pressure on the product designers to find creative ways to enhance product functionality at the target cost. The process concludes with component-level target costing, which creates intense pressure on the suppliers to undertake creative cost reduction.

The multifunctional nature of target costing is illustrated by the way finance, marketing, design engineers, production, and suppliers must interact to make the target costing process a success. The target selling prices that marketing establishes must be viewed as realistic, the target profit margins that finance sets must be considered achievable, and the strategic cost-reduction challenge and target cost-reduction objective must be perceived as outcomes of intelligent negotiations between product design and top management. Finally, the suppliers must accept the component-level target costs as reasonable selling prices that will enable them to make an adequate return. Only then will the target costing process be able to create the intense discipline required throughout the product design process to achieve the target costs.

CHAPTER 11

LESSONS FOR ADOPTERS

INTRODUCTION

Survival strategies in today's intensely competitive environments differ from those that were effective in the past. Japanese lean enterprises have learned that adopting the generic strategy of confrontation is often the only viable way to ensure corporate survival. At the heart of the confrontation strategy lies an integrated approach toward managing the survival triplet. Inherent to this approach is aggressive cost management that begins when the product is designed and that continues until it is discontinued. Firms that find themselves facing increased levels of competition should consider developing an integrated approach to cost management.

Before adopting target costing and value engineering, it is important to determine whether the firm's environment is changing in ways that support the adoption of the techniques and whether the benefits exceed the costs. Although simple diagnostic tests to determine the outcome of that trade-off don't exist, the firm should ask and answer the following six critical questions prior to adoption:

- Is profit management becoming more critical to your firm's survival?
- Is satisfying your customers becoming more critical to the survival of the firm?
- Is product design becoming more critical to your firm's survival?
- Are supplier relations becoming more critical to the survival of your firm?
- Is cost management the right place for your firm to expend resources?
- Can you create the right organizational context to support target costing and value engineering programs?

The first question explores the increasing importance of profit management to the survival of the firm. The next three questions relate to the environmental conditions surrounding the three major sections that make up the target costing process. They deal with the growing importance of customer satisfaction (which drives market-driven costing), product design (which drives product-level target costing), and supplier relations (which drive component-level target costing). The fifth question deals with the relative benefits that can be derived from cost management as opposed to other projects to increase profits. The last question relates to the ability to create the right organizational context to support the target costing and value engineering processes.

Answering these six questions will provide the reader with a sense of the advantages to his or her firm of adopting these two feedforward cost-management techniques. The questions are phrased in such a manner that an affirmative answer indicates increasing support for the adoption of the two techniques.

QUESTION ONE

Is profit management becoming more critical to your firm's survival ?

The objective of target costing is to help ensure that the firm achieves its long-term profit objectives. As competition increases, it becomes harder to achieve those objectives. Target costing and value engineering become more beneficial as the intensity of competition increases and, in particular, when that competition revolves around product functionality. Two fundamental questions indicate impending confrontation.

Is it becoming increasingly difficult to sustain your competitive advantage? Firms in confrontation are characterized by their excellent skill in satisfying customer requirements. Product offerings tend to converge as firms quickly learn how to imitate each other's products and as they hone their market analysis skills. The sustainable competitive advantages enjoyed by differentiators and cost leaders are now being replaced by a series of temporary competitive advantages. Survival zones become narrow because of the shrinking difference between what is achievable and what consumers expect along the price, quality, and functionality axes of the survival triplet. Target costing and value engineering systems play an increasingly important role under such circumstances because of their ability to integrate and prioritize among the three axes of the survival triplet.

Are your competitors competing more aggressively along all three axes of the survival triplet? Lean enterprises have the ability to offer products of higher quality and functionality at equal or lower prices than traditional firms. Western managers' response to this threat tends to be either a retreat to niche markets or a feverish attempt to improve along all three axes of the survival triplet simultaneously. Within a confrontational industry, however, one of the three axes of the survival triplet typically dominates the other two, and firms compete by rapidly improving their performance along this one axis. Failure to maintain approximate parity along the other two axes can be equally disastrous, however. Target costing and value engineering help managers integrate cost management with product functionality and quality and set the right priorities when expending resources on increased functionality or quality versus reduced cost.

QUESTION TWO

Is satisfying your customers becoming more critical to the survival of the firm?

In highly competitive environments it becomes more critical to develop products that satisfy the customer. Japanese firms have demonstrated that target costing and value engineering are effective tools for helping product designers develop products that satisfy customers without sacrificing profitability. The following questions will assist you in determining whether customer behavior is changing in ways that support the adoption of target costing and value engineering.

Are your customers becoming more sophisticated? As customers become more sophisticated, they learn to differentiate between competitive offerings on the basis of ever-smaller differences in price, functionality, and quality. This ability forces firms to develop similar product offerings, causing survival zones to narrow. At the same time, customer loyalty typically decreases. Consequently, customers will readily switch to a competitor if a firm's products do not satisfy their requirements. Target costing and value engineering are proven tools for increasing the likelihood that a firm's new product design process will be customer driven and ultimately profitable.

Are your customers placing increased pressure on your profits? By demanding products with increased functionality and quality, often at reduced prices, customers place pressure on the firm's ability to earn profits. For example, customers may implement their own target costing systems and create significant pressure on the firm to reduce its selling prices. As selling prices fall, profit margins will be squeezed, and aggressive, sophisticated cost management becomes imperative if historical profit margins are to be sustained.

Are your customers demanding an increase in the rate of introduction of functionality? As customers demand an ever-increasing rate of introduction of new product functionality, target costing and value engineering become more beneficial. By increasing the degree of integration of customer requirements and product design, these two techniques help product designers develop products that reflect the firm's best estimates of where customer expectations will be when the product is launched.

Are your customers' requirements staying reasonably predictable? The success of target costing depends in part on the ability to predict customer requirements. In particular, it depends on the accuracy with which future selling prices and sales volumes can be predicted. The ability to predict selling prices is critical because they form the starting point for the calculation of allowable and hence target costs. The ability to predict sales volumes is equally important because the sales volume of a product plays a primary role in determining the magnitude of scale economies and learning curve effects incorporated into the product-level target costs.

In addition, supplier costs are similarly influenced by predicted volumes. When selling prices are predicated on actual component sales volumes, it becomes more difficult to ensure that component-level target costs are achieved. Thus, as long as reasonably accurate selling prices and sales volumes can be predicted, target costing will be feasible and provide significant benefits. In markets where customer requirements are virtually unpredictable, such as the fashion industry, less formal approaches to cost management are probably preferable.

QUESTION THREE

Is product design becoming more critical to your firm's survival?

Product design plays an increasingly important role in firm survival when customers are highly sophisticated. The narrow survival zones that result increase the difficulty of ensuring that products satisfy the customers' demands. The firm can deliver products that are more closely aligned with customer requirements in several ways, all of which support the adoption of target costing and value engineering.

Are you continuously extending your product lines? As customers become more demanding, one typical response is to increase the number of products in the product line through both vertical and horizontal differentiation. The greater the number of products, the greater the product design challenge and the higher the likely benefits from target costing and value engineering.

Is time-to-market becoming more critical? As the level of competition increases, it often becomes critical to bring products to market on time. As the rate of product introduction increases, there is less time to correct any design flaws that lead to excessive costs. Therefore, the intense discipline of target costing becomes ever more valuable.

Are your product life cycles continuously decreasing? A second way to increase customer satisfaction is to accelerate the rate of introduction of new products. This strategy enables products to be designed that better match the customer's current requirements. Shorter product life cycles create additional pressures to manage costs because there is less time to earn the profits required to offset development costs. Under these circumstances, target costing and value engineering increase in importance for two major reasons. There are more products to be designed and there isn't enough time to correct any design errors that cause the product to cost too much.

Is your product strategy evolving toward an incremental rather than a breakthrough innovation approach? Target costing and value engineering are most beneficial for firms that practice an incremental strategy of product innovation. In such firms, each new product generation relies heavily on the technology of earlier generations. This multigenerational stability of technology allows historical databases to be developed. These databases can be used to predict future costs with a relatively high degree of accuracy. However, if the firm's strategy relies heavily on the application of breakthrough technologies, then cost estimation is more difficult and target costing and value engineering become less applicable.

Are your products becoming more complex? One of the ways firms try to increase customer satisfaction is by designing products with enhanced functionality. This strategy often leads to products that are more complex because they contain more components and newer technologies and require additional production steps. These changes support the adoption of target costing because they create more opportunities to reduce costs in the design stage.

Are you spending more on up-front investments? Target costing and value engineering are more beneficial to firms with new products that involve high up-front investment costs, such as manufacturers of automobiles and bulldozers, than to firms that can afford to flood the market with new product offerings, such as soft drink manufacturers. The firm with relatively inexpensive product launch costs can afford more misses because there is less at stake. In contrast, firms that sell products that consume considerable human and capital resources during the design phase require disciplined processes such as target costing and value engineering for weeding out products of limited market potential early in the development cycle.

Is the new-product design process in your firm becoming increasingly difficult to manage? If there is pressure to lessen the time-to-market for new products, then firms may need increased coordination among functional groups. Ambitious schedules often require formerly sequential steps in the design process to be performed concurrently. For example, marketing must coordinate more closely with R&D to understand the risks associated with not-yet-mature technology slated for inclusion in the next generation of products. Test and manufacturing plans may have to be developed and capital investments made prior to the availability of a new product's specifications and drawings.

Target costing and value engineering are useful in such environments for two reasons. First, the cross-functional communication fostered by the target costing process is essential to achieve the necessary coupling among the functions under conditions of increasing uncertainty. Second, target costing and value engineering systems guide a decision-making process that aligns the goals of the firm with the needs of the customer. This process can be decentralized so that unanticipated events can be resolved in the timely manner required to get products to market as quickly as possible.

QUESTION FOUR

Are supplier relations becoming more critical to the survival of your firm?

For many firms, the transition toward lean production includes the increased outsourcing of components and subassemblies that are of lesser strategic importance. Lean enterprises are characterized by a greater degree of horizontal integration than their mass production predecessors. Therefore, for lean firms, managing the supplier base becomes a even more critical survival skill. The following questions will help you determine if supplier relations are changing in ways that support the adoption of target costing and value engineering.

Is your firm becoming less vertically and more horizontally integrated? If the firm is buying a greater percentage of components from its external suppliers for each new generation of product, it is becoming de facto more dependent on them. As the degree of horizontal integration increases, target costing and value engineering potentially can provide more benefits because a greater proportion of the product's costs can be disciplined through component-level target costing. The degree of benefit depends on the relative power of the supplier-customer relationship.

Are you gaining power over your suppliers? Component-level target costing assumes the ability of the customer to set the selling prices of the components it purchases from suppliers. Target costing and value engineering will be most beneficial if the firm is large compared to its suppliers and can dictate selling prices to them. Consequently, a firm that is growing faster than its suppliers or is selecting smaller suppliers from which to source its components will get increasing benefits from adopting target costing and value engineering.

Can your supplier relationships be made more cooperative? Target costing and value engineering provide increased benefits when deployed across the supply chain. The ability to extend their implementation beyond the firm's boundaries depends largely on the level of trust and cooperation between the firm and its suppliers. The key to successful interorganizational cost management is first to identify a select group of suppliers capable of rapid innovation and aggressive cost management and then to reward these suppliers for their creativity while using target costing and value engineering as disciplining devices.

QUESTION FIVE

Is cost management the right place for your firm to expend resources?

It takes considerable resources to implement sophisticated cost management systems such as target costing and value engineering. Before implementing such systems, it is important to ensure that other projects are unlikely to provide greater returns. Some firms may be better off trying to increase sales instead. The anticipated profit margins play a significant role in determining where resources are best expended, as illustrated by a simple example.

Assume that firm A, which has effective patents and highly loyal customers, enjoys 70% profit margins and that firm B, which faces a confrontational environment, has only 7% profit margins. Suppose that for the same level of investment both firms can identify projects either to increase sales or decrease costs. Project 1 increases sales by 10% and project 2 decreases costs by 10%. Each project has an equal up-front cost but no on-going costs. Which type of project should each firm undertake? The answer depends on the level of profitability as illustrated in the following pro forma financial statements for the two firms, assuming each adopts project 1 or 2.

	Firm A	Firm B
Today's profitability		
Sales	$100	$100
COGS	30	93
Profit	$ 70	$ 7
Project 1: Increase sales by 10%		
Sales	$110	$110
COGS	33	102
Profit	$ 77[*]	$ 8
Project 2: Decrease costs by 10%		
Sales	$100	$100
COGS	27	84
Profit	$ 73	$ 16[*]

[*] This project reports the highest profits.

Firm A maximizes profits through increased sales, while Firm B, the one in confrontation, does so through cost reduction. This difference between the two firms highlights why cost management becomes more critical as competition becomes more intense and profit margins fall.

Even firms with relatively high profit margins that are facing increasing levels of competition should consider implementing cost management systems such as target costing and value engineering before margins are squeezed. It takes several years to develop the internal discipline and organizational culture necessary for such systems to be effective. For example, target costing and value engineering both demand a high level of integration among the product design, marketing, and manufacturing engineering functions. Achieving this integration by breaking down the functional silos of a traditional organization takes a considerable length of time. If the firm waits to take action until profit margins are low, they may discover that they have missed the train and lack the resources and time to implement the necessary systems.

QUESTION SIX

Can you create the right organizational context to support target costing and value engineering programs?

The organizational context in which a new cost management system is implemented is critical to its success. Many contextual factors should be considered when implementing a target costing and value engineering system. These factors include ensuring top management support, structuring the work force, identifying appropriate champions of the new system, and ensuring adequate support from the accounting and finance function.

Will top management at your firm provide the necessary support for the target costing and value engineering program? Top management must play an active role in supporting the implementation of target costing and value engineering. These persons must communicate the importance of cost management to the organization and reward its success. Critically, they must help develop an organizationwide culture of aggressive cost management. It is not sufficient for just a few individuals to be worried about managing costs—everyone in the organization must develop a cost-down mind-set.

For target costing and value engineering to be effective, the cost-reduction objectives must be achievable, especially when the techniques are first introduced. When the firm is facing competitors that are considerably more efficient, it may take several product generations to achieve competitive parity. Top management must be willing to accept lower profits while the firm catches up, instead of setting overly aggressive cost reductions that the engineers and design teams view as unachievable. The resulting violations of the cardinal rule would risk reducing the power of target costing through a degradation of its discipline. The solution is to accept that a significant strategic cost-reduction challenge exists and provide for a multigenerational plan to achieve the allowable cost. This plan would allow the target cost for the next two or three generations to be significantly above the allowable cost, but for each generation the challenge would be reduced until in the end it is reduced to minimal levels. The firm will not achieve its desired level of profitability during this period of adjustment, but by using a disciplined target costing system it will be improving relative to its competitors and will be able to compete aggressively in the future.

In general, the organization must be given time to develop the expertise necessary to take full advantage of target costing and value engineering. Implementing target costing requires a multiyear effort in which marketing, engineering, and manufacturing must learn to communicate more effectively and share power in new ways. Target costing and value engineering can create powerful strategic advantages over competitors that do not use them. But these techniques work through a process of continuous improvement, not through abrupt changes. While techniques such as reengineering can provide large immediate benefits, they often damage the culture of the firm in the process. Sometimes a gradual approach to change is more successful in the long run: a firm that can reduce its costs a few percentage points a year faster than its competitors will eventually dominate them (all else being equal).

Will the work force support the adoption of target costing and value engineering? Effective target costing and value engineering require a committed, motivated, and aware work force. Such a work force is critical to the success of all cost management techniques because it is the work force that is responsible for achieving the cost-reduction objectives. In Japanese firms, the work force is typically organized into self-guided teams that have considerable but conditional autonomy. As long as the teams achieve their objectives, they are given considerable latitude. If they fail to achieve them, however, their autonomy is reduced until they are back on target.

Every team is involved in cost management and has a cost-reduction target. In most firms, the procedure for setting cost-reduction targets for teams is part of a hierarchical target-setting process. The process begins with corporatewide cost-reduction targets set during the annual planning process. The corporatewide cost-reduction target is then distributed among the divisions. At this stage, the corporate planning department and the divisional managers negotiate over the targets. In the next stage, the divisional cost-reduction targets are distributed among the production facilities in the divisions and then to the teams.

In some firms the cost-reduction objectives are set from the bottom up and in others from the top down. Under either approach considerable negotiations are undertaken before the final cost-reduction objectives are set. These negotiations are critical if the groups are to commit to achieving their objectives. Imposed cost-reduction objectives rarely succeed.

The role these teams play in achieving the cost-management objectives is critical. Without these teams, the grassroots commitment to cost reduction would not occur. Firms thinking of implementing target costing and value engineering should analyze their organizational culture to see whether they have an appropriately structured and motivated work force.

Can the right project champions be identified? Experience with the implementation of other strategic cost-management techniques such as activity-based costing has highlighted the importance of having the appropriate champions for the analysis and action stages. In the analysis stage, the system is being designed and installed. The analysis champion is responsible for ensuring that the design team receives the resources and the support it requires from the various functional areas that are necessarily involved in the development of the target costing and value engineering system.

The action champion, who is usually different from the analysis champion, is the person who ensures that the system is used to create competitive advantage. It is not sufficient to follow a "field of dreams" strategy and build the system, simply assuming that the users will follow. Instead, a proactive approach is required that encourages users to apply the technique on specific projects the moment the system is available. It is the action champion who must help identify these projects and create a supportive environment in which the technique can be used successfully.[1]

Can you identify an appropriate pilot project? Like most technical innovations, target costing and value engineering may require

---

[1] For more on the role of champions, see Cooper et al, *Implementing Activity-Based Cost Management: Moving from Analysis to Action,* Montvale, NJ, Institute of Management Accountants, 1992.

a successful pilot project before top management will support wide-scale implementation. To prove the concept, the pilot project should:

- demonstrate the magnitude of savings possible with target costing and value engineering,
- be limited enough in scope to complete in a reasonable length of time, and
- be large enough to create sufficient visibility in the firm to justify wide-scale implementation.

As part of the pilot project, a cadre of people will have to be trained in target costing concepts and design principles. The firm's product designers may have to study value engineering techniques and begin developing the databases to support the target costing process. For example, cost and functional tables may have to be developed for the pilot project. In addition, the firm may have to adopt QFD and other techniques that support cost management in the product design process.

Even though it might be possible to complete the pilot within the existing organization structure, the pilot design team should emulate the organizational structure the firm will adopt under target costing. For example, a chief engineer will have to be identified and given total responsibility for the project. In addition, the various functions will have to name members of the multifunctional design teams and give them the necessary support. Finally, the team must adopt the philosophy of continuous cost reduction that underlies target costing and develop the common language it requires.

Will the accounting and finance function support the project? The role of the accounting or finance department is critical. Again, lessons learned from the implementation of activity-based costing systems shed light on the appropriate role for management accountants in implementing target costing and value engineering systems. The accounting department must play a supporting role, not a leadership role. The users are the ones who lead the process and become committed to the success of the system (i.e., its use). An accountant with the appropriate power can be the analysis champion but typically not the action champion. The action champion is usually a senior manager of one of the functional areas (marketing, engineering, or production) who has the ability to influence the design process significantly.

SUMMARY

The above questions are designed to help you determine if your firm should consider adopting target costing and value engineering. You should take six major themes into account—the importance of profit management, the difficulty of satisfying customers, the importance of product design, the nature of supplier relations, the relative importance of cost management, and finally, the ability to create the right organizational context. The way each question was posed helps you assess whether your firm's conditions are changing in a direction that supports the adoption of the two feedforward cost-management techniques. To help you gauge the relative position of your firm, Table 11-1 indicates the supporting direction for each condition identified as being relevant to the adoption decision.

Firms facing competitive environments that predominantly support the adoption of target costing and value engineering should not hesitate. Target costing and value engineering provide considerable payoffs to early adopters. The firm that can most rapidly reduce its costs without having to compromise product quality and functionality will develop a powerful competitive advantage that will continue until its competitors follow suit.

TABLE 11-1. SUPPORTIVE DIRECTION FOR EACH FACTOR

Critical Parameters	Supporting Direction	Your Firm
Importance of Profit Management		
Intensity of competition along all three axes of the survival triplet	Increasing	
Ability to sustain competitive advantage	Decreasing	
Profit margins	Decreasing	
Difficulty of Satisfying Customers		
Ability of customers to differentiate between competitive offerings	Increasing	
Customer loyalty	Decreasing	
Ability of customers to impose target prices	Increasing	
Customer expectations about the rate of introduction of new product functionality	Increasing	
Ability to predict selling prices	High	
Ability to predict sales volumes	High	
Importance of Product Design		
Number of products in the product line	Increasing	
Time-to-market	Decreasing	
Product life cycles	Decreasing	
Incremental as opposed to breakthrough innovation	Increasing	
Product complexity	Increasing	
Up-front investments	Increasing	
Complexity of new product design process	Increasing	
Nature of Supplier Relations		
Degree of horizontal integration	Increasing	
Relative power over suppliers	Increasing	
Ability to make supplier relations cooperative	Increasing	
Relative Importance of Cost Management	Increasing	
Ability to Create the Right Organizational Context		
Management's willingness to support the implementation of target costing and value engineering	High	
Management's willingness to develop an organizationwide culture of aggressive cost management	High	
Management's willingness to set achievable cost-reduction objectives and accept lower profits in early years of adoption	High	
Management's willingness to give the organization time to develop the necessary expertise	High	
Motivation and commitment of the work force	High	
Willingness of the entire organization to get involved in cost management	High	
Likelihood of identifying appropriate target costing and value engineering champions	High	
Ability to identify a pilot project and design team with the right characteristics and skills	High	
Willingness of the accounting and finance function to play a supportive role in the implementation process	High	

PART 3

THE CASES

CHAPTER 12

NISSAN MOTOR COMPANY, LTD.

INTRODUCTION

Nissan Motor Company, Ltd. (Nissan) was by 1990 the world's fourth-largest automobile manufacturer. In 1990, Nissan produced just over 3 million vehicles, supplying approximately 10% of the world's demand for cars and trucks. Of these vehicles, slightly over 2 million were passenger cars. Nissan, founded in 1933, considered itself the most highly globalized of the Japanese automobile companies, producing vehicles at 36 plants in 22 countries and marketing in 150 countries through 390 distributorships and over 10,000 dealerships.

Nissan had a stated policy of increasing its globalization through a five-step process:

- increasing local production,

This case was prepared by Professor Robin Cooper of the Peter F. Drucker School of Management at Claremont Graduate University. The assistance of Ms. May Makuda, KPMG Peat Marwick, is gratefully acknowledged.

- raising the local content of its products through expanded use of locally sourced parts and components,
- strengthening local research and development capabilities,
- localizing management functions, and
- localizing decision-making processes.

As a result, four of the five major overseas manufacturing plants were managed by local chief executive officers, and in 1990 regional headquarters were opened in Europe and North America.

The domestic Japanese passenger automobile market was intensely competitive. The largest manufacturer was Toyota, with approximately 45% of the domestic market. Nissan was second with approximately 25%, followed by Honda and Mazda, which together represented about 20%. In an attempt to increase its market share in the expanding but fiercely contested domestic market, Nissan implemented a plan to achieve annual domestic sales of 1.5 million cars by 1992 and to obtain the number-one rating in terms of customer satisfaction. This strategy depended on designing products engineered around clearly defined concepts chosen to offer customers automobiles that matched their lifestyles.

Automobile firms had been steadily increasing their range of products since the 1950s. This increase was driven primarily by changes in consumer preferences. For example, U.S. consumers in the 1950s viewed the automobile as a status symbol, using the make of automobile they owned to signal the level of their economic success. Nissan executives characterized the consumer in this era as "keeping up with the Joneses." As automobile ownership became more widespread, consumers began to view their automobiles as making a statement about *who* they were. Nissan executives characterized the consumer of this era as "doing his or her own thing." Consumer demand of the 1960s required more variations and a broader range of model types than in the 1950s. As the 1960s closed and individualism became less important, consumers came to view their automobiles as making statements about *what* they were. Nissan executives characterized this era as being dominated by a desire for a "consistency of lifestyle," that is, bankers wanted automobiles appropriate for bankers. This shift required that the production of more models and variations to satisfy consumer demand. During the 1980s, consumers started to demand automobiles that suited multiple lifestyles. As one executive summed up the transition, "the old segmentation that assumed a single lifestyle no longer worked; we now have to design cars that allow people to be bankers by day and punk rockers at night." This transition in consumer preference placed additional pressure on the firm to increase its range of product offerings.

Despite this pressure, Nissan had chosen to reduce systematically the number of distinct models it would introduce in the 1990s. This decision reflected two trends. First, the differences between consumers in the three major markets—Japan, North America, and Europe—were decreasing, and second, the costs associated with introducing new models were increasing. The decrease in differences among consumers in the three major markets reduced the need to develop models specific to a single market. The increased costs associated with launching new models made it difficult to make acceptable profits if too many new models were introduced each year. These trends suggested to Nissan top management that overall profitability would be increased by reducing the number of distinct models supported, while maintaining the same level of effort to design and market the remaining models.

Introducing New Products

Over the years, Nissan had developed a formal procedure to introduce new products. One of the major elements of this procedure was a sophisticated target costing system. In this system, a target selling price for each new model was first established. Then a target profit margin was determined based on corporate profitability objectives. Finally, the model's target cost was identified as the difference between the target selling price and the target profit margin. Once the target cost of the new model was established, value engineering was used to ensure that when the new model entered production, it could be manufactured at the desired target cost.

The procedure to introduce new models had three distinct stages. In the conceptual design stage projects to introduce new product models were initiated. In the product development stage, the new models were readied for production. In the production stage, they were manufactured. Taken together, these three stages lasted about 10 years, the average life cycle of a modern passenger automobile. The conceptual design stage required about two years to complete, the product development stage required about four years, and the production stage typically lasted about another four years. Thus, it was not unusual for Nissan to be simultaneously producing the current model, preparing its replacement for introduction, and conceptualizing that model's replacement.

The Conceptual Design Stage

New product models were designed in the conceptual design stage. First, the designers identified the mixture of models that Nissan expected to sell over the next 10 years. This mix was described in a matrix of vehicles by major market and body type (e.g., coupe or sedan). The matrix contained qualitative information about each model, such as its price range, target customers and their income levels, and the range of body types supported. This information was maintained for both current and future models and effectively described each model's market position. The primary purpose of the product matrix was to ensure that Nissan achieved the desired level of market coverage.

The firm identified new entries in the matrix using consumer analysis. This analysis was undertaken by market consulting firms using a number of different techniques, including general economic, psychological, and anthropological surveys as well as direct observation. In recent years, this analysis had identified over 50 potential models that theoretically could be introduced successfully by Nissan. However, top management had identified the optimum number of models that Nissan could successfully support at under 30. This number was limited by several factors, including the cost of differentiating each model in the minds of consumers, research and development, and the cash flow associated with maintaining dealer floor inventory. Thus, the challenge that Nissan management faced was to select the approximately 30 models that would maximize market coverage.

New models were conceptualized by identifying consumer mind-sets. Mind-sets captured characteristics of the way consumers viewed themselves in relation to their cars. These mind-sets could be used to identify design attributes that consumers took into account when purchasing a new car. Typical mind-sets included value seeker, confident and sophisticated, aggressive enthusiast, and budget/speed star. By identifying clusters of these mind-sets, Nissan could identify niches that contained a sufficient percentage of the automobile-purchasing public to warrant introducing a model tailored specifically for that niche. For example, the Sentra (a mid-price family sedan) was designed to satisfy the confident/sophisticated and value seeker mind-sets, while the ZX (a high-performance sports car) was designed to satisfy the budget/speed star and aggressive enthusiast mind-sets. As a Nissan marketing executive commented, "If we believe that a sufficient market will exist in four to five years, then we will develop a model to fit it." Thus, each model with its body shape variations—sedan, coupe, hatchback, and wagon— was designed specifically to satisfy a different group of consumers.

As a final check on the appropriateness of the proposed model lineup, each model was categorized using three primary attributes: performance, aesthetics, and comfort. For example, comfort was considered the most important attribute for the Sentra model and performance the least important, while for the ZX model, performance was considered the most important and comfort the least. A plot of the attribute characteristics of both current, future, and competitive models allowed top management to determine that the proposed product mix covered an adequate percentage of the market.

At this stage of the product introduction process, the conceptual design was sufficiently developed to allow a rough estimate of the number of vehicles to be sold and the costs associated with its development. These estimates were used in a life-cycle contribution study to estimate the overall profitability of the proposed model. The purpose of this study was to ensure that the new model was likely to generate a positive contribution over its life. The life-cycle contribution study was a comparison of the estimated revenues generated by the new model to the expected cost of the product across its life (see Exhibit 12-1). The model's revenue was determined by using a rough estimate of the selling price and the anticipated volume of sales. The anticipated direct material cost, which included raw material, paint, and purchased parts, was subtracted from this estimate. The difference between these two quantities was the estimated direct material marginal profit of the new model.

From this profit, four additional sets of expenses were subtracted. The first set contained the direct manufacturing and sales expenses, predominantly driven by the number of units produced and sold. The direct manufacturing expenses included elements such as the cost of the energy, cutting tools, and indirect materials consumed. The direct sales expenses included logistics costs such as shipping and delivery costs. The second set contained the estimated direct labor costs. The third set contained the depreciation charges for machining, die casting, and other major production steps. Finally, the estimated research and development expenses were subtracted to give the life-cycle contribution of the particular model under development.

The depreciation charges used in the life-cycle contribution analysis were estimated by taking the total depreciation for each machine or

EXHIBIT 12-1. LIFE-CYCLE CONTRIBUTION STUDY

+ Selling price
− Direct material cost [a]

= **Direct material marginal profit**

− Direct manufacturing expenses [b]
− Direct labor
− Depreciation [c]
− Research and development [d]

= **Life-cycle contribution**

[a] These costs include raw material, purchased parts, and paint.
[b] These expenses include direct sales expenses.
[c] Depreciation is charged on machinery, dies, and conveyor lines.
[d] Excludes all capital items for which depreciation is charged.

process and dividing it by the number of units expected to be produced on that equipment over its life. If the equipment was dedicated to the new model, the typical case with stamping dies, then the number of units was the estimated volume of production for that model. If the equipment was common to several models, often the case with conveyors, then the number of units was the total of all units expected to be produced of all models using that equipment. The depreciation charge used for the life-cycle contribution analysis was not the one used for financial reporting purposes. Nissan reported depreciation using a declining balance approach for both tax and financial reporting purposes. However, for the life-cycle contribution calculation it used a straight-line approach. Management modified the depreciation calculation because it felt that the straight-line approach better captured the relationship between asset use and models produced than the declining balance approach. If the life-cycle contribution was deemed satisfactory, the conceptual design process was allowed to continue.

As the conceptual design of the new model progressed, additional consumer analysis and financial analysis were undertaken. The company used consumer analysis to obtain a better idea of the price range over which the model would sell and the level of functionality the consumer expected. The financial analysis consisted of a rough profitability study in which the profitability of the highest volume variant of the new model was estimated using historical cost estimates and the latest estimate of that variant's target selling price. This target selling price was determined by taking into account a number of internal and external factors. The internal factors included the position of the model in the matrix and the strategic and profitability objectives of top management for that model. The external factors considered included the corporation's image and level of customer loyalty in the model's niche, the expected quality level and functionality of the model compared to competitive offerings, the model's expected market share, and finally, the expected price of competitive models.

The first stage of value engineering was designed to determine whether the new model could be manufactured at an acceptable profit. The process began by developing an order sheet detailing the characteristics of the 20 to 30 major functions of the proposed model. Examples of the major functions included the engine, air conditioner, transmission, and sound system. The characteristics of each major function were chosen to satisfy the collection of consumer mind-sets for which the model was designed. For example, the engine specified for a ZX would be a high-performance one, while for the Sentra it would be smaller, less powerful, and less expensive. The current cost of the model was determined by summing the current manufacturing cost of each major function of the new model. This current cost was compared with the model's allowable cost to determine the level of cost reduction required to achieve the desired level of profitability.

The allowable cost of the new model was determined by subtracting its target profit margin from its target selling price. The target profit margin was determined by careful consideration of available information on the consumer, the firm's anticipated future product mix, and its long-term profit objective. Each new model's target profit margin was established by running simulations of the firm's overall profitability for the next 10 years if it were selling the models identified in the product matrix at expected sales volumes. The simulations started by plotting the actual profit margins of existing products. The desired profitability of planned models was then added (see Figure 12-1) and the firm's overall profitability determined over the years at various sales levels. This predicted overall profitability was compared to the firm's long-term profitability objectives set by senior management. Once a satisfactory future product matrix was established that achieved the firm's profit objective, the target profit margins for each new model were set.

To help minimize the risk that Nissan would not achieve its overall profitability targets, the simulations explored the impact on overall profitability of different price/margin curves for different product mixes. For example, historically higher margins had been earned on higher priced vehicles. However, with the reduced product offering and the increased profitability expected, the future curve might be higher. Alternatively, because there was no guarantee that the existing relationship between selling price and profit margin would remain unchanged, simulations were also

FIGURE 12-1. IDENTIFYING THE TARGET PROFIT MARGIN

run to explore the impact of fundamentally different relationships between selling price and profit margins.

The first stage of value engineering and the identification of the target selling price was an interactive process. When the allowable costs were considered too far below the estimated cost, the appropriate price range and functionality were reviewed until an allowable cost that was considered achievable was identified.

The excess of the current manufacturing cost over the allowable cost determined the level of cost reduction that value engineering had to identify. For example, the current manufacturing cost of the model might be ¥3,000,000 and the allowed cost ¥2,700,000, which is a required cost reduction of 10%. The next step in the value engineering process was to identify the allowable cost of each major function. This cost was set by teams derived from almost every functional area of the firm, including product design, engineering, purchasing, production engineering, manufacturing, and parts supply. Although the allowable cost was usually lower than the current cost, sometimes the allowable cost was higher because the new product specifications demanded higher performance and functionality than existing designs. In total, the sum of the cost reduction for each major component was meant to equal the required level of cost reduction to achieve the model's allowable cost.

Several critical decisions about the model were made during this stage of the conceptual design process, including the number of body variations, the number of engine types, and the basic technology used in the vehicle. For example, the original concept for the model might include a five-door variant. However, if during this stage of the analysis it was determined that developing such a variant would be too costly or take an excessive amount of time, plans for a five-door variant would be postponed to the next version of that automobile. Once the projected cost of each major component had been identified, the expected cost of manufacture could be computed.

After the first value engineering stage was completed, the firm conducted a major review of the new model. This review included an updated profitability study and an analysis of the performance characteristics of the model. In the profitability study, the expected profitability of the model given by the target selling price minus the target cost was compared to the latest estimates of the capital investment and remaining research and development expenditures required to complete the design of the product and allow production to commence. In the performance analysis, factors such as the quality of the hardware, engine capacity, exhaust emissions, and safety were considered. If both the financial and performance analyses were considered acceptable, the project to introduce the new vehicle was authorized, and the model shifted from the conceptual design to the product development stage.

The Product Development Stage

The first step in the product development stage was to prepare a detailed order sheet listing all the components required in the new model and to analyze it to see which components would likely be sourced internally versus externally. Suppliers, both internal and external, were provided with a description of each component and their potential production volumes. Suppliers were expected to provide price and delivery timing estimates for all the components.

The next step in the development of a new model was to produce the engineering drawings for trial production. Value engineering was used at this stage of product development to determine allowable costs for each of the components in every major function of the automobile. This estimate was achieved by identifying a cost-reduction objective for each component.

Cost-reduction objectives for components were identified in several ways. First, competitors' products were purchased, disassembled, and analyzed. This analysis sometimes generated ideas for cost reduction.

Second, the firm used an incentive plan to motivate the parts suppliers to generate cost-reduction ideas. If an idea was accepted, the supplier that suggested the cost-reduction idea would be awarded a significant percentage of the contract for that component for a specified time period, say 50% for 12 months. This incentive scheme was viewed as particularly important because even if a cost reduction could not be achieved for this model, it signaled to the suppliers that when the next model was developed this component would be subject to cost-reduction pressures.

Management then identified ways to increase the commonality of parts across variations and models (e.g., the same seats might be used in two different models) and to reduce the number of components in each model. Originally, for example, plastic nuts held in place the kick plates used to protect the door. To eliminate the nuts, a way was developed to mold the plastic interior of the door so that no nuts were required.

To avoid having to develop target costs for all 20,000 components in a typical new model line, the engineers performed detailed target costing on only two or three representative variations. Each variation contained approximately 3,500 components, and typically 80% of the components were common across variations. As a result, about 5,000 components were subjected to detailed target costing. The target costs of the other 15,000 components were estimated by comparing them to similar components in the 5,000 already target costed. The completion of this target costing exercise provided cost-reduction objectives for all the components in the new model. The comparison of the allowable cost of each function and the sum of the expected cost of the components in that function after cost reduction indicated whether the major function could be produced at about the allowable cost. When the sum of the component costs was too high, additional cost reductions were identified until the total target cost of the representative variation was acceptable. The target costs for each component were compared to the prices quoted by the suppliers. If the quoted prices were at the right level, the quote was accepted. If the initial quote was too high, then further negotiations were undertaken until an agreement could be reached.

The next phase in product development was to construct two or three prototype vehicles. The construction of these prototypes was instructive. First, any components that were difficult to assemble were identified. Typically, these components or the assemblies into which they fitted were redesigned to improve the ease of assembly. Second, assembly times could now be estimated quite accurately. In the third stage of value engineering, the effect of these redesigns on the target costs of the components was determined and assembly target costs were identified. The output of this stage of value engineering was called the final target cost. It differed from the draft target cost in two ways. First, it included assembly costs and second, the indirect manufacturing costs. The indirect manufacturing costs were assigned to products using the same procedures as the firm's cost system. Thus, the final target cost for a model variant was expected to be equal to its reported product cost during manufacture.

A comparison of the final target costs for each model variant and its expected selling price allowed the anticipated profit on the vehicle to be determined. Marketing reviewed the expected selling price in light of recent competitive products and market conditions and recommended a final selling price. Accounting was responsible for authorizing the actual selling price of each variation. It took into account marketing's review of existing conditions, the final target cost of the vehicle, and the target profit margin for the vehicle. Accounting notified marketing of the recommended selling price, once it was set. In Japan, this was the price at which the car would be sold across its life. In other markets such as North America, incentive plans and other marketing techniques could cause the effective price to change across the life of the product.

Accounting was not involved in the value engineering process, which was the responsibility of the cost design and engineering department. The primary function of accounting was to set the final target cost for each model variant and ensure that the vehicles were manufactured for that amount. As the vehicle entered production, accounting would monitor all component and assembly costs, and if they were not in line with the final target costs, accounting would notify cost design and engineering that the final target costs were not being met. When the target costs were exceeded, additional value engineering was performed to reduce costs to the target levels. Thus, the fourth and final value engineering stage ensured that the actual component and assembly costs were equal to their final target costs.

Unless the production cost exceeded the target cost, no cost-reduction efforts were undertaken during the production stage. Management had determined that the incremental savings from such efforts were more than offset by disturbances they created to the production process. When inflation or other factors caused costs to rise, pressure was exerted on the suppliers to find ways to keep component costs at their final target levels. Similarly, pressure was exerted on the assembly plants to achieve the assembly target costs.

The Production Stage

The Zama facility, located a few miles from Tokyo, was one of Nissan's five major domestic manufacturing facilities. Built in 1964, it was 852,000 square meters in area and contained two complete stamping and assembly facilities. In addition, the facility housed a car delivery area and the firm's machinery design center. No production parts were produced at Zama. The plant was involved only in producing pressed metal parts, welding them together, assembling the body, painting it, and then assembling the finished automobile.

The Zama plant was designed to produce 90 cars per hour, operating on a two-shift basis with production occurring for 15 hours and 20 minutes per day. A two-shift operation enabled the plant to produce between 1,300 to 1,400 cars per day at full capacity. While preventative maintenance was carried on throughout the day, primary maintenance was performed during the 8 hours and 40 minutes in which no production was scheduled. Zama was a highly automated plant. Of the approximately 3,000 spot-welds per car, over 97% were performed automatically. This high level of automation had been achieved over a number of years. The current high level of automatic spot welding was achieved around 1980. To sustain this high level of automation, Zama contained nearly 300 robots. All these robots were designed by Nissan, but only about 40% were actually manufactured by the firm.

In 1990, the facility was dedicated to the production of two models and three body types. The Sunny, or Sentra as it was called in the North American market, was produced in two body types: the four-door and two-door coupe configurations and the Presea in a single four-door configuration. The facility could produce each body type in numerous variations of key components, such as engine, air conditioning, and transmission. When all possible option variations were included, Zama produced approximately 20,000 different variations of each of the three distinct body types.

The large number of variations forced the facility to produce cars to customer order. This production strategy fit well with Nissan's corporate strategy of providing customer satisfaction, high quality, short delivery times, and high functionality. In fact, fast delivery was considered so important that the production strategy was called by a name that translated to "deliver the car with the paint still wet." In 1990, a car ordered from a Japanese dealer and produced in the Zama plant could be delivered to the customer within two weeks. Because in Tokyo it required at least a week to get the certificate of space required to enable a car to be purchased, the effective wait for a new Nissan was negligible. This short delivery time despite high product diversity was achieved via aggressive use of just-in-time production.

The Product Cost System

Nissan used the same cost system throughout its assembly plants. The system reported full product costs that included both direct and indirect expenses. It traced the indirect expenses to the products in two different ways. Direct and indirect manufacturing expenses were charged directly to the production cost center in which they were consumed and then allocated to the products. Service and administrative expenses and corporate expenses were allocated to the product without first being allocated to a production center. Corporate expenses were equal to about 15% of the sales revenue. They consisted of three major types of expenses: product-related expenses that included advertisements, warranty, and delivery expenses; geographic-related expenses, such as the costs associated with the sales division in Tokyo that supported sales in all three major markets; and finally, the expenses associated with corporate administration, legal, and accounting. The product-related portion of corporate expenses was approximately 30%; the geographic-related portion, 50%; and the other expenses, about 20%.

The cost system calculated three different product profitabilities. The first was the direct material marginal profit. It was calculated by subtracting the cost of the raw materials and purchased parts from the selling price.

The second was the product contribution, determined by subtracting direct manufacturing costs, research and development expenses, and corporate expenses directly related to the product from the direct material marginal profit. Direct manufacturing expenses included manufacturing supplies, such as cutting tools, and machine depreciation. The procedures used to assign these expenses to products varied depending on the nature of the expenses and how they were consumed. For example, the costs of manufacturing supplies were assigned to the products based on the number of direct labor hours the product consumed in the center if the department produced several different models, and number of units if only one model was produced. Machine depreciation was assigned to products based on the number of units irrespective of model. Research and development expenses included the cost of the labor, facilities, and supplies consumed in the research and development facility. They were allocated to products based on the labor hours consumed in research and development on that product. The corporate expenses directly related to the product included items such as product advertisements, product incentives, warranty costs, and delivery expenses.

The third product profitability, operating profit, was calculated by subtracting indirect manufacturing, service, administrative, and the remaining corporate expenses from the product contributions. Indirect manufacturing expenses were directly charged to the production cost center in which they were incurred. Examples of indirect manufacturing expenses included transportation, maintenance, and facility depreciation. These expenses were assigned to the products based on the total direct cost of the products produced in that center. Service and administrative expenses were assigned to the products without first being assigned to the production centers. These expenses were assigned based on the total direct cost of the product. Corporate expenses were first assigned to the facility and then to the products based on their total direct costs in the same way as were service and administrative expenses. The breakdown of product profitability is shown in Exhibit 12-2.

The Nissan cost system was undergoing continuous modification. In particular, a program had been initiated to trace as many costs as possible directly to the production departments. Thus, over time the service and administrative categories were dropping in relationship to the direct and indirect manufacturing costs. The success of this program and the heavy reliance on external parts and service suppliers were visible from the relative importance of the three major cost categories. The direct costs, that is, those costs that were traced directly to the products, represented 85% of total manufacturing costs. The direct and indirect manufacturing costs that were charged to the production departments and then assigned to the products represented about 10% of total manufacturing costs. Service and administrative costs amounted to only 5% of total costs.

The product costs reported by the cost system had four primary uses:

- in the long-range strategic plan as a basis for estimating future profitability,
- for cost-control purposes, in particular to ensure that across the production life of a product its target cost was maintained,
- to help select the product mix, in particular with respect to the variations of a given model,
- to identify unprofitable variants that were candidates for discontinuance.

EXHIBIT 12-2. BREAKDOWN OF PRODUCT PROFITABILITY

Revenue   ****

Direct material cost
   1. purchased parts   ****
   2. raw materials   ****   (****)

Direct **Direct material marginal profit**   ****
cost and
expenses Direct manufacturing cost
   1. direct labor cost   ****
   2. manufacturing expenses directly variable to the
(Directly   number of units produced (e.g., cutting tools)   ****
characterized   3. tools and machinery   depreciation charge   ****
to products)   maintenance cost   ****
                 energy cost   ****   (****)

   R&D expense (labor, expenses, facilities)   (****)

   Corporate expenses directly related to product
   1. product advertisements   ****
   2. product incentives   ****
   3. warranty   ****
   4. delivery expenses   ****   (****)

Product contribution   ****

Indirect manufacturing cost consumed by
manufacturing shop floors
   1. indirect materials (e.g., monkey wrench, sandpaper)   ****
Indirect 2. indirect labor cost   ****
cost   3. shop floor facilities   depreciation charge   ****
                 maintenance cost   ****
                 energy cost   ****   (****)

Services
   1. quality inspection (labor, expenses, facilities)   ****
(Irrespective 2. logistics (labor, expenses, facilities)   ****
of product   3. energy consumed in service department   ****
characteristics) 4. maintenance consumed in service department   ****   (****)

Administration
   1. labor cost   ****
   2. general expenses   ****
   3. facilities   depreciation charge   ****
            maintenance cost   ****
            energy cost   ****   (****)

Corporate expense (excluding directly related to product)
Corporate 1. expense consumed by geographic sales division
expenses   a. maintenance costs of sales network   ****
     b. labor cost consumed in the divisions   ****
     c. general expenses consumed in the divisions   ****
   2. general administration   ****   (****)

Operating profit   ****

CHAPTER 13

TOYOTA MOTOR CORPORATION

If Toyota does one thing better than other automakers, it is cost management. After earning a reputation for quality and fuel efficiency in the "economy models," we moved successfully into upscale models, like the Lexus line. But we still pride ourselves in the cost competitiveness of our products in every price stratum. The history of Toyota is a story of unceasing efforts to reduce costs.

1992 Annual Report

INTRODUCTION

Toyota Motor Corporation (Toyota) started as a subsidiary of the Toyoda Automatic Loom Works, Ltd. It was founded in 1937 as the Toyota Motor Company, Ltd. It changed its name to the Toyota Motor Corporation in 1982 when the parent company merged with Toyota Sales Company, Ltd.

This case was prepared by Professor Robin Cooper of the Peter F. Drucker School of Management at Claremont Graduate University and Professor Takao Tanaka of Tokyo Keizai University.

In 1993, Toyota Motor Corporation was Japan's largest automobile company. It controlled approximately 45% of the domestic market. Its next largest Japanese competitor was Nissan, with approximately 25% market share, followed by Honda and Mazda, which together represented about 20%. The remaining 10% of the domestic automobile market was made up of several domestic manufacturers, including Isuzu, and several foreign competitors, such as Mercedes Benz and the "big three" American firms: General Motors, Ford, and Chrysler.

The domestic and world automobile markets were characterized by intense competition. Models were brought out rapidly despite their high development costs. Fractions of a percentage of market share frequently were viewed as representing the difference between success and failure.

GLOBALIZATION

Over the years, Toyota had changed from a Japanese firm into a global one. In 1993, a considerable part of the firm's overseas markets were serviced by local subsidiaries that frequently designed and manufactured automobiles for local markets. For example, in North America local plants produced almost one-third of the vehicles sold in that market. These vehicles were produced in three plants. One was in Kentucky, another was in Ontario, Canada, and the third was the New United Motor Manufacturing Inc. (NUMMI) joint venture plant with General Motors.

Between them these plants produced approximately 400,000 vehicles per annum, including 220,000 Camrys and 170,000 Corollas, the remainder being pickup trucks. Production volumes for pickup trucks were expected to increase to approximately 100,000 in the next few years. In 1994, the firm expected to begin exporting vehicles from North America to markets such as Japan and Taiwan. In addition to automobiles, the firm also manufactured and sold forklifts. Toyota controlled 70% of the forklift market in the United States.

The same commitment to local manufacture and control was apparent in the firm's other major overseas markets. In Europe, the two new U.K. plants began producing engines and passenger cars at the end of 1992. Unit production of cars was expected to reach 100,000 by 1995 and 200,000 units before the end of the century. Altogether, Toyota vehicles were either manufactured or assembled in more than 20 nations. These local manufacturing facilities provided jobs for nationals and business for local supplier firms.

The relative importance of the international supplier business to Toyota was increasing. In 1992, for example, Toyota purchased locally approximately 70% of its parts requirements (or $5 billion) for its North American operations. The other 30% was imported from Japan, but this percentage was expected to decrease over time. Worldwide, by 1994 Toyota expected to purchase $6.3 billion of parts from local suppliers and import $2.9 billion for domestic use.

Product design was also international in scope. Calty Research, Inc., a Toyota subsidiary formed in California in October 1973, was responsible for the body styling and interiors of new models scheduled for production in North America. The design styling for European markets was coordinated from the firm's design and technical centers located in Brussels.

SUPPLIER RELATIONSHIPS

Third-party suppliers were responsible for approximately 70% of the parts and materials required to produce the automobiles manufactured by the firm. This high level of dependency on externally supplied items made supplier relations extremely critical to the firm's success. In particular, the cost and quality of third-party-supplied parts was considered critical. In recent years, the expansion of the firm into international production had required increased interaction with non-Japanese suppliers. A primary focus of these interactions was to raise the efficiency and quality of these suppliers to the same level as Toyota's suppliers in Japan.

To help non-Japanese supplier firms manufacture acceptable parts, Toyota had developed programs to transfer Japanese manufacturing techniques. At the heart of these so-called design-in programs was joint work by suppliers and Toyota engineers on new components. This joint work commenced in the early stages of the vehicle-development process. These programs were developed because prospective suppliers cited a lack of involvement in the early stages of vehicle design as an obstacle to winning business in high-value components. The design-in programs were intended to lessen this barrier and allow new suppliers to succeed. A typical design-in program consisted of a competition between several designers for the contract for a specific part. The competing firms were evaluated based on the prices bid, the technology applied, and their performance. Toyota granted the winning firm a contract for the life of the model. When the next model was developed the contract was once again thrown open for bidding.

By 1993, more than 120 U.S. suppliers had participated in design-in programs and successfully managed to design parts acceptable to Toyota. Another 200 firms were going through such programs but had yet to sign contracts for parts. A similar program was in place in Europe.

Another way that Toyota supported its overseas suppliers was to help them adopt the Toyota Production System. The Toyota Production System contained four key elements: just-in-time (JIT) production, kanban, total quality management, and multifunctional work teams. The advantage of just-in-time production was that it avoided the build up of excessive work-in-process inventories and increased the firm's ability to respond quickly to customer demands. Kanban acted as the driving force behind JIT by tying production closely to customer demand. Total quality management was necessary to ensure high-quality products and to minimize the risk that the reduced levels of inventories would result in stockouts caused by poor-quality components. Finally, multifunctional workers capable of performing several tasks were needed to deal with the increased complexity of the production process. To assist Toyota's overseas suppliers in adopting these production methods, Toyota engineers spent time helping them convert to the Toyota approach. Overall, this program had been successful, with many Toyota overseas suppliers implementing modified versions of the Toyota Production System.

COST PLANNING

Cost planning at Toyota was primarily a program to reduce product costs at the design stage. Toyota first set its cost planning goals and then set out to achieve those goals through aggressive design changes. To assess correctly the gains made, the exact amount of cost reduction achieved through design changes was estimated after excluding all other factors that affected costs, such as increases in material and labor prices.

The measurement process used cost tables to estimate the current cost of existing models. These cost tables were kept up-to-date for changes in material prices, labor rates, and production volume levels and helped determine both depreciation and overhead charges. The estimated cost of the existing model was used as the basis for estimating the cost of the new model without additional savings. Comparison of this estimated cost to the vehicle's target cost gave the desired level of savings, or cost-planning goal, as it was called.

At the profit-estimation stage, also referred to as the "target cost-setting stage," Toyota calculated the differences between the costs of the new and current models, distributed the appropriate portion of the cost-reduction goal to the design divisions, and then assessed the results.

Profit targets for the life of the new model were also calculated as differences between estimates and targets. Setting goals and assessing the results based on cost differences between old and new models constituted the essence of budget control at Toyota. The idea was that cost reduction goals for each control unit should be specified clearly to ensure attainment of the overall goals of the company.

TARGET COSTING

In 1959 Toyota invented target costing as it has developed in Japan. Although many major manufacturers in Japan use target costing, the system used at Toyota Motor Corporation is the oldest and considered by many the most technically advanced. While the idea of systematic cost reduction had existed at Toyota since it was founded, the process was first codified in the mid-1960s, when the firm set itself the objective of producing a $1,000 car.

Cost estimation played a role in target costing, but the two differ in several ways. Cost estimates relied on existing standards while target costs were adjusted for any future savings due to design changes. Also, cost estimates had a horizon of six months while the horizon for target costs was the time remaining until the launch of the new product.

The primary use of target costing was to bring the target cost and the estimated cost of a product into line by better specification and design. Simply estimating the cost of new products was not the purpose of the target costing system. Its ultimate goal was to enable a product to attain its profit targets throughout its life.

PRODUCT PLANNING

There were two broad categories of product development at Toyota. One was for completely new types of automobiles and the other was for changes to existing models. The development of an entirely new model, such as the Lexus, was relatively unusual. In contrast, projects to modify existing models were normal. In Japan, passenger cars usually underwent major model changes every four years, but recent industry trends suggest that the period between full model changes may become longer in the future.

Toyota used target costing primarily to support model changes, although it followed the same general cost control procedures for the design of entirely new vehicles. The primary difference between the two types of projects was the level of uncertainty in the cost estimates, which were much higher for projects involving new types of automobiles.

A model change began with a proposal for development of a new model. Chief engineers usually made these plans. The new model plans included specifications such as size (length, width, wheelbase, and interior space); weight, mileage, engine (type, displacement, and maximum power); transmission (gear and moderation ratios); chassis (suspension and brake types); and body components. It also included the development budget and development schedule and the retail price and sales targets.

New models basically maintained the same product concept as their predecessors. The development plan might define some specifications for the new model, but styling was left unspecified; usually the plan mentioned no more than a vague image. Most of the cost incurred in any model change was for prototyping. The level of prototyping costs increased in proportion to the number of test models built.

Retail Prices and Sales Targets

The sales divisions usually proposed retail prices and sales targets. The fundamental guiding principle for setting the retail price was that the price should remain the same unless there was a change in function from the previous to the new model, and this change altered the perceived value of the vehicle in the eyes of the customer. In theory, therefore, prices changed in accordance with the perceived value of the vehicle.

Increases in retail price were based primarily on the customer's perception of additional value from new functions (e.g., the introduction of four-wheel steering and active suspension) or from better performance (e.g., higher engine horsepower or better fuel efficiency).

Formula for List Price of a New Model

Toyota viewed the selling price of a new car model as being made up of the selling price of the equivalent existing model plus any incremental value due to improved functionality. For example, adding air conditioning to the standard version of a model would increase its price by the value of air conditioning as perceived by the customers. The incremental value of a new model was determined by analyzing market conditions. Due to the maturity of the automobile industry, most new features already existed in some form on other models. For example, if air conditioning was to be included in the standard version, its added value was determined using the list price of optional air conditioners for other models. If no equivalent option existed, a rare event, then the firm's design engineers and market specialists would estimate how much customers were likely to be willing to pay for the added feature.

The price increase for an added function was not always equal to its selling price as a stand-alone option. The incremental price for a given increase in functionality might be lowered because of the firm's strategy for the vehicle model in question and because of the pricing strategies of competitors. As functions were added to the standard version the selling price rose until it reached the upper limit for that class of vehicle. This upper limit was the maximum selling price that the firm believed it could set for the new vehicle. When this limit was reached, the only potential benefit from adding functionality was in increased sales.

The uncertainty associated with market conditions when the product was to be introduced some four years after the design project began forced the firm to delay setting the functionality of the standard version as long as possible. Therefore, the target price and margin for the product (thereby the associated target unit price) *were set quite some time before product launch.*

The exact functionality of the standard version was set only when factors such as competitive offerings, foreign exchange rates, and user demand were better understood. Changing the functionality of the standard version allowed Toyota to increase the probability that the new model would achieve its desired level of profitability. Similarly, the actual selling price was not fixed until just before the product was launched. Delaying these two critical decisions reduced significantly the uncertainty faced by the firm. For example, the incremental value assigned to an air bag in the U.S. market might have been $450, but the competition had set the incremental value at $700. In this case, Toyota might increase its price by the difference. Similarly, if the competitive prices were lower, Toyota would drop its prices to match.

The sales division proposed anticipated production volumes based on past sales levels, market trends, and competitors' product offerings. The sales division typically proposed a figure that was considered safe, that is, achievable. This safe figure was based on the model's current sales level. Optimism was restrained in favor of realistic goals.

With the help of engineers in the design, test-production, and technical divisions, a chief engineer drafted the development plan for the new model and then led the development project. Well over 100 engineers from the various divisions worked with a chief engineer on a typical project, but since they belonged to different divisions not all team members were under the chief engineer's supervision. Probably only about a dozen people reported directly to him. In this sense, the chief engineer was more a project leader than a supervisor of product development. This approach had several advantages. First, the chief engineers were responsible for coordinating the design process at the design divisions.

The design divisions were relatively autonomous, and chief engineers were expected to develop a "concept" for the new vehicle that spanned multiple design divisions. Keeping the design divisions autonomous was considered important, as it allowed them to use their expertise across multiple projects. The tensions created by this matrix approach were considered beneficial to the creative design process and worth any conflict they created.

Cost Planning

The cost-planning goal was set based on the product plan and the targets for the product's retail price and production volume. Given that an estimated price had been established, setting this cost-planning goal was equivalent to setting the product's target profit: the expected profit from the total sales of the product over its production life, which was usually four years. The product's target cost was the unit cost on which the profit target was based.

Calculating Target Profit and Target Cost

The lifetime target profit for a product such as the Celica was calculated by multiplying the target sales volume by the model's return on sales or, as it was known at Toyota, profit ratio of sales. The sales profit ratio was set with reference to the corporation's long-term target profit ratio.

Estimated cost was the cost of the new product determined by using the firm's cost tables. An estimated profit was calculated using this figure. Estimated profit was less than the target profit because the target cost included the estimated cost savings due to value engineering and other cost-reduction activities. The difference between target and estimated profit was the amount to be cut from costs through cost planning. The cost-planning goal was obtained by subtracting the estimated total profits from the target profits.

The goal of cost planning was to determine the unit profit needed to achieve the profit target and thus the amount to be trimmed from the new product's cost through cost-planning activities. Estimated profit was equal to the retail price minus the estimated cost per unit times the production volume. As cost-reduction activities were implemented, the product's estimated costs decreased. If the goal was achieved, the target cost and expected cost became equal, as did the expected and target profits.

Estimating Difference Costs

Rather than adding all the costs for a new model, Toyota's unique approach to cost planning was to sum the differences in cost between new and current models. This approach had several advantages. Cost planning could begin even before blueprints for the first test model were drawn. Second, estimating the total difference instead of the total cost tended to be less troublesome and more accurate, and finally, it helped the related divisions understand cost fluctuations. The differences approach was considered more accurate because the typical new model was heavily based on existing designs. Trying to estimate the cost of a new vehicle from scratch would, in management's opinion, introduce more errors than using existing information and modifying it accordingly. The approach was more helpful to the design divisions because it highlighted the areas of the new model that were different from existing designs. It was these new designs that required most of the work in the design divisions.

The estimated cost of a new model was therefore described as the cost of the current model plus the cost of any design change. Thus, for every increment in the functionality of a new model there was an estimated incremental price and cost. This approach allowed the firm to measure the incremental profitability of each new function it built into a new model.

A full model change necessarily involved many differences in design. Consequently the cost of the design change was broken out into the costs of a number of different design modifications. Estimating differences, in addition to the advantages already mentioned, helped clarify the cost-planning goal and showed accurately how much was accomplished through cost planning.

The main concern of cost planning was the design of the new model. Its effectiveness was measured as the amount of cost reduction achieved through design. Other factors that affected cost, including wages and fluctuations in indirect costs incurred by related divisions, had to be eliminated from overall cost reduction to identify the portion due to cost planning. By fixing the cost of the current model and calculating the differences between the current and new models, Toyota's system dealt only with cost changes resulting from changes in design and production volume.

Without actual drawings for the new model, the estimate often began with just an idea. Rough sketches provided by the design division were often the only sources of information. Since the people at the design and cost-planning divisions had the latest information on the results of basic in-house research, they were better qualified than the accounting division to estimate the costs from the sketches. Design divisions were responsible for the design of each major function of the new vehicle. Major functions included the engine, transmission, air conditioner, and audio system. There were about 20 design divisions at Toyota.

Applying the results of basic research to product design was helpful both for improving product performance and for attaining the cost-planning goal. Cost planning tested the company's design capabilities. If the new model had the same functionality as existing models, then the initial cost estimates were assumed to be unchanged.

Equipment Investment and Depreciation

A full model change inevitably required a large investment in equipment. The amount invested in terms of depreciation affected the cost of the new model. For this reason, an outline of equipment investment was often provided at the product-planning stage, but the budget for equipment investment was not officially set until after the product plan was approved.

Equipment investment was divided into two categories: investments in equipment needed to replace metal molds and investments to add, expand, or improve other equipment and facilities. When new parts or systems were added, such as four-wheel steering, production lines were built to manufacture them. When production methods altered because of a model change, the lines were updated. If a plant did not have enough capacity to make the new model, it needed to expand.

Many automobile parts (e.g., presswork, sheet metal, plastics, and castings) were made using metal molds that were costly to manufacture. A model change required alterations in many body and interior parts, forcing the firm to acquire new metal molds for those parts whether they were manufactured in-house or by parts suppliers. These molds were integral parts of the production machinery so they were considered production equipment.

The production-technology division developed the equipment-investment budget. Since this budget covered investments required for the production of the new car, the need to undertake the investments was not questioned, but the specific figures in the budget were not always accepted outright and sometimes had to be modified. The investment budget might be reduced if the reviewing committee believed that more efficient use of common facilities was possible or that the same level of productivity could be achieved with less expensive machinery. Finally, the accounting division produced an adjusted plan based on the budget proposed by the production-technology division. This plan took into consideration the influence on cost of the new model and the balance between the budget and the company's total equipment investment.

Promoting Cost Planning

The purpose of cost planning was to determine the amount by which costs could be reduced through better design of the new model. The cost-planning goal was distributed to the divisions in charge of design for the model, with each division being given a particular cost-reduction target. For example, the divisions in charge of design of the engine, body, chassis, drive train, electronics, and interior all received different cost-reduction targets.

Distributing Cost Targets

Toyota management believed that it was impossible to attain a target cost simply by mandating a uniform reduction of costs for all divisions. Consequently, the chief engineer distributed a portion of the goal to each design division. Discussion continued with each division until both the division and the chief engineer were satisfied with the amount, which was based on precedent and experience. The divisions were responsible for attaining their cost-reduction goals.

The chief engineer was expected to make his own decisions about where cost reduction was to occur. One of the objectives of the target costing system was to focus the attention of the design engineers in the design divisions in the right place. The chief engineer typically had objectives for the new vehicle that affected where costs could be reduced. For example, he might want a quieter car or a higher-performance engine. To achieve these objectives, he would decrease the magnitude of the cost reductions expected from the design divisions responsible for those aspects of the product and increase the expected reductions from other divisions.

The cost-reduction targets were first assigned to the design divisions and then to the part level for certain major parts. For example, if one of the design changes was to increase the power of the engine by 10 hp, the engine division might estimate that the improvement would increase the cost of the engine by ¥X. The chief engineer would use the precedents for upgrading engines by 10 hp to estimate a more aggressive cost and then ask the division to compromise on ¥Y. In contrast, another division might be asked to reduce costs because the new part would be smaller or lighter than the old one. A third division might be asked to maintain the same cost despite a change in materials because no change in performance was anticipated.

Design Policy and Cost

There were more opportunities for cost reduction during product planning than during the actual development stage. These opportunities varied, depending on the specific stage of product planning. The turning point was when styling was determined and production of the first prototype was about to begin. Decisions before this point had more effect on cost than those taken after this point. For example, the Celica line consists of four sister versions—the Celica, the Corona Exiv, the Carina ED, and Curren. Toyota planned the 1990 models to be mass-produced at the rate of 7,000-10,000 units per month. About 3,000 Celica units were designated for the Japanese market; the rest, more than twice that number, were for export. The versions look quite different, but they have much in common in the engine and chassis. Differentiating the versions while using as many common parts as possible was vitally important to the product lineup.

Determining Components

Before going into details of design, the components for each car type were identified. The number of different parts required by the new model and the resulting total equipment investment change depended heavily on the degree to which parts were shared across the different versions of the automobile. Generally, mass-produced parts as opposed to small-lot parts reduced cost. If the versions shared common parts, cost could be calculated based on production volumes of 20,000 units per month. If each used certain parts specific to it, the cost would be calculated based on a volume of 10,000 units. If parts were common across other product lines, additional cost savings were typically achieved.

Body styling also had a major influence on cost. Some designs created complex part structures and required higher tolerances, thus increasing the number of labor-hours needed for production. For example, in the early 1990s a trend emerged to make the bumper look as if it was an integral part of the body. The space between bumper and body was reduced from several millimeters to less than a millimeter on recent models to achieve this objective. This change inevitably increased the required tolerances, thus increasing manufacturing costs. When a certain body style required a cost increase, the chief engineer had to decide how large an increase was acceptable.

VALUE ENGINEERING

Value engineering began once the cost-planning goals—the amounts of cost to be cut—were distributed to the design division. The designers' top priority was to create high-quality, high-performance products that satisfied the customer. At the same time, they were expected to attain their cost targets.

Each design division became responsible for attaining its respective cost-reduction goal, but the specifics of parts, materials, and machining processes were left to its discretion. The chief engineer made exceptions for large, especially costly parts. He would sometimes specify cost-reduction targets for specific parts to the related divisions. These specific-part cost-reduction targets were set at the same time as the divisional targets. For example, consider a part estimated to cost ¥3,000. If it was judged that a cost break on this particular part would contribute significantly to attaining the target goal for the entire model, the chief engineer might ask the related design division for a part-specific cost reduction of perhaps ¥500.

The design divisions often organized value engineering meetings to help attain their cost goals. Since test parts were developed at about this time, value engineering could be based on the prototype. Three test models were usually made, which meant that the cycle of drawings, part production, and value engineering was repeated three times over a period of about one year. The design was complete when the performance and cost goals were attained. The final mass-production plan was then established.

Various Value Engineering Methods

There were various methods for conducting value engineering. It generally started with performance checks on the test parts. Designs were changed to give the parts their specified performance, neither more nor less. Then discussion turned to possible ways to cut costs while maintaining performance. There were no formulas or manuals for value engineering, but there were several areas where it was possible, including material specifications and consumption, yield, number of parts, ease of work, and work hours. For instance, a part that used too many fasteners would be redesigned so that it would use fewer fasteners. A design would be changed if a projection showed that a change in shape would make production easier. Special parts would be replaced with mass-produced ones if the final performance would be unchanged.

Value engineering was undertaken on the basis that a "cost savings of a yen here and a yen there eventually mounts up." Even small savings were identified and achieved on the grounds that 10 such changes would add up to a significant saving, even though each individual saving was relatively small. Concentrating effort on expensive parts and parts for which cost had increased markedly sometimes worked to increase the effectiveness of the value engineering process. The objective was not to change the target cost of the product but to increase the savings achieved from this area of the product.

KNOWING THE EFFECT
OF DESIGN CHANGES

To be effective designers must know how design affects such things as material consumption, yield, machining methods, and line time. The best designers were ineffective if they were not fully conversant with production techniques. Design engineers often lacked hands-on production experience. They were, therefore, expected to work closely with the production divisions to build their practical design skills.

People assigned from the cost planning and accounting divisions (the cost-planning group) were responsible for accurately gauging changes in cost following design modifications. There were about 100 such people in the two divisions. Several were assigned to each product line to support the design engineers. These estimators provided designers with such information as the effect on cost of a change in the machining process and the cost per minute of machine time.

Cost estimators used cost tables to calculate unit prices for manufacturing. Cost tables were developed for five major production steps: machining, coating, body assembly, forging, and general assembly. Cost tables detailed the machine rates for each step in the production process. These rates included labor, electricity, supplies, and depreciation costs. The exact form of the cost table rate depended on the type of production step being analyzed; for example, for stamping, the cost table contained the cost per stroke while for machining it contained the cost per machine-hour. Cost tables were highly detailed, and each production line had its own cost table identifying the cost per stroke of each press. Rather than the basic costs used for budget management, a cost table used for cost planning showed cost per production line, which was manufacturing costs broken down into direct labor costs and indirect line costs. In standard costing, shops or cost centers were groups of two or more lines.

Value engineering was not purely intellectual work that a design engineer could do at a desk, working from drawings. Instead, value engineering was based on efforts to improve production on the shop floor. Value engineering was not effective if production was not sufficiently well organized.

FROM COST PLANNING
TO MASS PRODUCTION

Since the main concern of cost planning was design, cost planning was effectively finished when the project entered the mass-production stage. Unless something unusual happened, Toyota rarely failed to attain its cost-planning goals. Follow-up studies were undertaken for about a year after start-up to ensure that mass production was going forward at the planned standard cost. The standard cost and target cost for a product were not identical. The target cost was established before all the details of production were known, so it didn't reflect the current conditions, such as the going rate for labor and materials or the specific plant in which the vehicle would be manufactured. In contrast, the product's standard cost was adjusted for all these factors.

The essential point was that target cost in cost planning and standard cost for mass production were treated as different standards with different functions. In cost planning, costs were estimated from the cost tables as the sum of the differences between the current and new models, though at that point the planners did not know on which lines production of the model would occur.

Standard cost at the mass-production stage changed depending on the specific production line on which the product was manufactured and the prevailing conditions at the time. For instance, production on lines working below capacity pushed costs up, while production on lines working at close to full capacity led to the best cost performance. At the cost-planning stage, it was difficult to imagine the details of line conditions for every part and thus to reflect these conditions accurately in cost estimates. At the mass-production stage, the lines that worked best under the current circumstances were chosen for production of the new model. ("Best" in this context means at the optimum for the entire company, which was not always optimal for the new model.) Once the lines were chosen based on these criteria, the standard cost was calculated. The production division then began its effort to maintain or even improve on the standard cost. Such improvements were under the auspices of the firm's kaizen program.

CHAPTER 14

KOMATSU, LTD.

INTRODUCTION

Komatsu, Ltd. is one of Japan's largest heavy industrial manufacturers. Founded in 1917 as part of the Takeuchi Mining Co., Komatsu Ironworks separated from its parent in 1921 to become Komatsu, Ltd. By 1991 Komatsu was a large international firm with revenues of ¥989 billion and net income of ¥31 billion. The company was organized along three major lines of business: construction equipment, industrial machinery, and electronic-applied products. Together these three lines of business generated about 80% of corporate revenues. Other operations, which accounted for the remaining 20% of corporate revenues, included construction, unit housing, chemicals and plastics, and software development. Construction equipment and industrial machinery were considered core businesses while electronics-applied products and other operations were considered new businesses.

This case was prepared by Professor Robin Cooper of the Peter F. Drucker School of Management at Claremont Graduate University.

Copyright © 1994 by the President and Fellows of Harvard College. Harvard Business School case 194-037.

In 1989, the company adopted a "3G" strategy of *growth, globalization*, and *group diversification*. The growth objective required all divisions to expand aggressively, with 1995 sales expected to reach ¥1.4 trillion. The globalization objective was to achieve worldwide production by the year 2000. In 1993, the firm's equipment was used in over 160 countries and was manufactured on three continents in 11 countries. The group diversification objective sought to aggressively develop three new business areas: electronics, plastics, and robotics. By the year 2000, the firm expected all nonconstruction products, including these three areas, to account for 50% of group revenues.

CONSTRUCTION EQUIPMENT

Komatsu was the world's second-largest manufacturer of a complete line of construction equipment. The firm's product line contained more than 300 models, including bulldozers, hydraulic excavators, wheel loaders, and dump trucks. With a more than 30% share of the domestic excavator market, Komatsu was the largest player in the Japanese market. Four other major players in the excavator market were Hitachi, with just under 30% of the market; Kobelco, with about 15%; Caterpillar Mitsubishi, with about 12.5%; and Sumitomo, with under 10%. Only Komatsu and Caterpillar Mitsubishi produced both bulldozers and excavators. The other three firms produced only excavators.

The number of competitors in the excavator market reflected the large market for those products in Japan. In the early 1990s, Japan represented over 50% of the world market for excavators, due to the mix of construction projects in that country. Most Japanese construction projects were in urban settings and were relatively small in size. Excavators were more practical than bulldozers for such applications because they were more versatile and less expensive. Excavators could perform applications such as digging, carrying, moving, and loading dirt, while bulldozers could only move dirt.

PRODUCT DEVELOPMENT AND DESIGN FOR MANUFACTURABILITY COST STUDIES

The product development process at Komatsu lasted two years on average. If the redesign was relatively minor, the process might take as little as six months, while complete redesigns might take as long as three years. The process contained four major stages: product planning, design, trial production (including testing), and preparation for full production.

The Product Planning Stage

The product planning process, last updated in 1981, began with the preparation of a long-range development plan. This plan, prepared at the same time as the firm's long-range production and sales plans, described the mix of products that Komatsu expected to sell over the next five to 10 years. The plan described the functionality of products that had yet to be designed in conceptual terms only. These conceptual designs, consisting of detailed descriptions of the structure of the major subassemblies, were developed before the products entered the design stage.

The Design Stage

The purpose of the design stage of product development was to prepare the product for prototype production. The design phase consisted of three major steps: a conceptual drawing of the product, layout drawings for the product, and detailed parts drawings.

During the first step, the first of four design-for-manufacturability studies (the A study) was conducted. The A study evaluated the feasibility of achieving the target cost and the overall manufacturability of the design. (Exhibit 14-1 outlines the major objectives of the A study.) The A study culminated in a meeting of the product manager, the design managers, and the production managers of the plants at which the product would be produced. If concerns about the product's design were voiced at this meeting, further analysis of its design was undertaken. Once the product passed through this evaluation stage, layout drawings were produced.

EXHIBIT 14-1. PURPOSE OF THE A STUDY

- Evaluation of prospects to achieve the cost target (planning and coordination department and purchasing department)
- Evaluation of number of basic specifications and fastening capability of attachments (planning and coordination department of each plant)
- Evaluation of manufacturing possibility through current facilities and current techniques and confirmation of incorporating production technology and research outcomes into products (manufacturing departments)
- Evaluation of timing required for implementation of new techniques and new facilities (manufacturing departments)
- Evaluation of whether main components and parts should be manufactured in-house or through outside contractors (manufacturing department and purchasing department)
- Evaluation of problems associated with purchasing parts (purchasing department)
- Evaluation of problems with the transportation of products (purchasing department)
- Evaluation of periods required for solution of the two preceding problems (purchasing department)

EXHIBIT 14-2. PURPOSE OF THE B STUDY

- Follow up A study manufacturability study results (various related departments)
- Evaluation of interchangeability and common ratios of use (planning and coordination department of each plant)
- Evaluation of VE improvement plan (planning and coordination department of each plant, manufacturing departments, and purchasing department)
- Evaluation of work lines for main parts (manufacturing departments)
- Extraction of items to improve process capability in producing main components and main parts (manufacturing departments and purchasing department)
- Evaluation of time required for mass-production preparation (manufacturing departments)
- Selection of suppliers/subcontractors (purchasing department)

The layout drawings contained a more detailed description of the product and its subassemblies. After the layout drawings were complete the second, or B, study was the next step. This study evaluated in more detail the firm's ability to achieve the target cost. (Exhibit 14-2 outlines the major objectives of the B study.)

Once the product passed this hurdle, the preparation of the parts drawings was approved. The parts drawings allowed the third, or C, study to be undertaken. Its purpose was to ensure that the product could indeed be built at the target cost and to confirm the actual manufacturing process and the facilities at which the product would be manufactured. (Exhibit 14-3 outlines the major objectives of the C study.) After the C study, a meeting of the development committee authorized prototype production. This committee consisted of the directors of the three technical divisions, the executive managing director of the corporation, the managing directors of both the domestic and overseas sales divisions, and the managers of related departments.

The Trial Production Stage

The objective of the trial production stage was to finalize the design of the product. Trial production made prototypes of each product, and was

EXHIBIT 14-3. PURPOSE OF THE C STUDY

- Extraction of manufacturing-process improvement items (manufacturing departments and purchasing department)
- Evaluation of facilitating measures for work and assembly (manufacturing departments and purchasing department)
- Confirmation of specifications for purchasing parts (purchasing department)
- Extraction of mass-production preparation items (manufacturing departments and purchasing department)

Exhibit 14-4. Purpose of the D Study

Manufacturing departments and purchasing department shall, during the processes of trial-manufacture and quality confirmation, evaluate the process capability and other necessary matters and, if there is any discrepancy, issue a trial-manufacture problem document.

conducted at the plants at which the product was to be manufactured. The completion of trial production allowed the final, or D, study to be carried out. (Exhibit 14-4 summarizes the major objectives of the D study.) This study examined the ease of production and assembly of the new product and confirmed its quality. If any problems were encountered, a trial-manufacture problem document was issued. This document identified all the problems that had to be resolved before mass production could commence (e.g., design changes to improve the ease of manufacturability and reduce assembly time). Expected production costs were reestimated at this time. After trial production was completed, the prototypes went through a series of comprehensive tests designed to ensure the quality and durability of the final product. Any problems encountered during this testing phase were fixed before the product went into full production.

The Preparation for Full Production Stage

Once the product achieved the desired quality targets, the marketing committee received results of the trial production, estimates of final target costs, and various studies so they could give final approval to move into mass production. The marketing committee was the most senior committee in the firm. It consisted of the president, the executive managing director, the managing directors of the technical and sales divisions, the plant managers, and the product manager. Once approved by this committee, the product was readied for full production. The production drawings were prepared and the preproduction plans were developed. As part of the preproduction plans, any potential problems identified in the D study were confirmed and rectified. After this step, the product was released for mass production.

Reducing the Time to Market

Players in the highly competitive market for excavators and bulldozers had begun to compete on the basis of the time it took to get new products to market. Part of Komatsu's plan to improve its design for manufacturing was to change its relationship with its suppliers. In 1993, Komatsu manufactured about 30% of its products, designed and subcontracted another

50%, and purchased the remaining 20% from outside suppliers. The firm set target costs for the subcomponents manufactured by its suppliers and expected its suppliers to find ways to achieve these targets.

Though target costs were supposed to be negotiated with suppliers, Komatsu management was concerned that in reality these negotiations were relatively one-sided. In addition, management felt that the suppliers were brought into the negotiations too late in the design process. To allow the suppliers to have greater input into the design process, Komatsu initiated periodic meetings between the suppliers' research and development staff and its own research and development staff. The aim of these meetings was to integrate the research and development efforts of the two groups, give suppliers a chance to provide input much earlier in the design process, and ensure that target cost negotiations were more substantive.

TARGET COSTING

Throughout the product development process, target costs played a critical role in ensuring that the product would be profitable when released for mass production. The preliminary target costs used in the long-range development plan for major subassemblies, such as the engine, power train, and cooling system, were developed from prior experience with similar subassemblies and discussions with production and engineering.

These preliminary target costs or target values, as they were known, were used to help decide when cost-reduction techniques should be applied. Komatsu used three different cost-reduction techniques: design analysis, functional analysis, and productivity analysis. Design analysis identified the approximate structure of the major subassemblies in new products; the other two techniques were used to decide on target costs for the subassemblies.

Design analysis involved creating alternative designs for major subassemblies and selecting between them. Once the design approach for the major subassemblies was chosen, their target costs were determined using either functional or productivity analysis, depending on who was responsible for designing the subassembly. Functional analysis, a procedure for deciding on the target cost of a subassembly based on its functional characteristics, was used for parts designed and produced outside of Komatsu (such as cooling systems, hydraulic devices, and electrical subassemblies) because this procedure did not rely on detailed knowledge of the production process. Productivity analysis, a procedure for ascertaining the target cost of a subcomponent based on its manufacturing process, required more in-depth knowledge of the production process. It was therefore used for subassemblies designed by Komatsu, such as vehicle main frames, buckets, and gears, and manufactured either by Komatsu or by one of its subcontractors.

Design Analysis

Product engineers were expected to create several alternative designs for each major subassembly of a new product. Two factors were taken into account when choosing between these alternatives: quality and cost. A new design was adopted only if it achieved the desired levels of both quality and cost. Frequently, one of the proposed designs was for a higher-quality but more costly product, and the product engineers would explore ways to manufacture it at a lower cost. If they could find a cost-effective way to implement the new design, it was adopted; otherwise, it was abandoned or subjected to further study for future applications.

The process of design analysis at Komatsu can be illustrated by a change in the way the engine and torque converter, transmission, and steering clutch and brakes were positioned in the firm's larger bulldozers. In the old design, these three modules were physically separate. This approach required 86 hours to mount and dismount these modules during maintenance. Komatsu's customers had indicated that this mount/dismount time was a critical factor when they selected a bulldozer.

Design analysis identified two different ways to position the three components. The first approach integrated the three modules into two. One consisted of the engine and torque converter and the other, the transmission and steering clutch and brake modules. The integration of the latter two into a single module reduced the mount/dismount time to 44 hours. The second approach also integrated the three modules into two, but this time the torque converter, transmission, steering clutch, and brakes were integrated into a single module. This approach had the advantage of removing the need to change the oil, thereby reducing the mount/dismount time to 33 hours. Unfortunately, both the new design alternatives were more expensive than the old design; the faster new design was also the more expensive one.

This conflict between quality and cost was resolved by changing the way the ripper mounting bracket was attached to the bulldozer. Bulldozers were used for a number of tasks, and changeable attachments increased their versatility. For example, rippers were used for breaking up hard surfaces while dozers were used for removing loose material. The ripper mounting bracket enabled the ripper to be attached to the main frame. The new approach allowed the mounting bracket to be welded instead of bolted to the main frame. Welding was cheaper than bolting, and the savings equaled the additional cost of adopting the alternative design of the engine, transmission, and torque converter.

Welding, while less expensive than bolting, required that every bull-dozer have a mounting bracket. Previously, Komatsu was able to sell bulldozers with or without an attached mounting bracket. The effective cost savings from adopting the new attachment approach and welding the mounting bracket thus depended on the percentage of bulldozers ordered with mounting brackets. This mounting ratio varied depending on the size of the bulldozer: the larger the bulldozer, the higher the ratio. When the mounting ratio was taken into account, the new approach was cheaper for large bulldozers but more expensive for small ones. Consequently, the new configuration of the engine, transmission, and converter, and the welded ripper mounting bracket approach were adopted for large bulldozers but not for small ones.

Functional Analysis

The process of functional analysis at Komatsu can be illustrated by the development of the target cost of an excavator cooling system. The process began with an analysis of the functions of the cooling system and how they were achieved. The primary function of the cooling system was its cooling capacity; secondary functions included how quickly it started cooling after the engine was first switched on and how stable a temperature it maintained. The determinants of cooling capacity were ranked in order of importance. The most important was the surface area of the radiator. The second most important was the size of the fan, followed by the rotation speed of the fan, the volume of water in the system, and the ambient air temperature.

The functional analysis began by plotting the cooling capacity versus the radiator surface area for all existing products that used the same type of cooling system. This information was maintained in functional tables. From this plot the average and minimum lines for existing equipment were determined. The average line was found using linear regression, and the minimum line was drawn so that it passed through the most efficient cooling systems. The required cooling capacity for the new model was used to identify the minimum cooling area needed according to the best designs. The minimum cooling area was the one that generated the desired cooling capacity on the minimum line (see Figure 14-1).

The target cost for the cooling system was determined by a similar process, using surface area versus cost information for existing products. This information was maintained in functional cost tables. A graph of the cost of cooling systems against their surface area was plotted for all existing models using the same cooling technology. The average cost per surface area was determined using linear regression. The minimum cost line was again drawn passing through the most cost efficient designs. The mini-

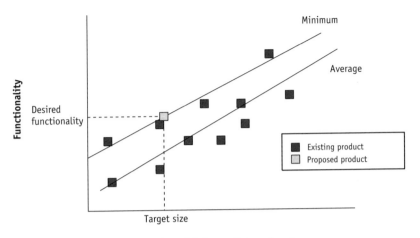

Size of Primary Determinant

FIGURE 14-1. COOLING CAPACITY VERSUS RADIATOR CAPACITY

mum surface area identified from the cooling capacity/radiator surface area analysis was used to identify the minimum cost of the new cooling system. This minimum cost for the minimum radiator surface area became the target cost for the radiator (see Figure 14-2).

The same techniques were used to generate target costs for the other major components of the cooling system. For example, the target cost of

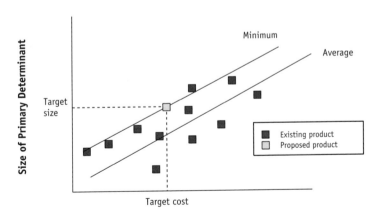

Cost of Primary Determinant

FIGURE 14-2. RADIATOR CAPACITY VERSUS COST

the fan in the cooling system was determined by plotting the size of fan against cooling capacity, identifying the average and minimum lines for the relation between fan size and cooling capacity. The minimum fan size-to-cooling capacity line was used to identify the minimum fan size. A plot of the cost of fans versus fan size was used to determine the minimum cost/size line and hence the target cost of the fan. This process was repeated for all major components of the cooling system.

The functional analysis approach changed for components such as shafts, where the function could be approximated by a simple physical measure such as the weight of the component. The function of the shaft was to connect the rotating output of the transmission to the wheels; the weight of the shaft identified the torque that it was able to handle. A shaft weight/cost table was used to develop a plot of the weight of shafts in existing products against their cost. From this plot, the average and minimum shaft weight/cost lines were identified and hence the target cost of the shaft. The target cost was the shaft cost predicted by the minimum cost line.

Productivity Analysis

Productivity analysis was used for major subassemblies designed by Komatsu. At the heart of this analysis was a set of tables that identified the cost of each production step as a function of its physical characteristics. For example, information was maintained about the cost of each type of material, the weight required by each design, the cost per meter of welding, and the length of weld required by each subassembly.

Productivity analysis analyzed the major steps in the production process of the new subassembly and compared the sum of their costs to the subassemblies' target cost. If the expected cost was too high, the section leaders responsible for each step in the production process had to choose a cost-reduction target for each step.

Ultimate responsibility for these cost-reduction targets lay with the product manager, who was responsible for ensuring that the new product successfully entered production. If the initial aggregated cost reductions were insufficient to allow the subassembly to be manufactured for its target cost, then the product manager and the production staff negotiated to increase the expected productivity savings. The final aggregation of the negotiated cost-reduction targets provided the latest estimate of the subassemblies' target cost.

The process of productivity analysis at Komatsu can be illustrated by the redesign of a mounting socket in the main frames of the firm's bulldozers. In the old design, the mounting socket consisted of a hole drilled through the body of the main frame. This design was simple to manufacture but had the drawback of creating a stress zone around the hole. To ensure that the mounting socket was strong enough, that section of the main frame had to be manufactured from expensive, high-grade materials. Productivity analysis had identified the reduction of the level of high-grade material in the main frame as one way to reduce costs. The new design called for welding a mounting bracket containing the mounting socket hole to the main frame of the vehicle. The new mounting unit was designed to reduce the strain imposed on the main frame so that normal-grade steel could be used.

CHAPTER 15

OLYMPUS OPTICAL COMPANY, LTD.: COST MANAGEMENT FOR SHORT LIFE-CYCLE PRODUCTS

INTRODUCTION

Olympus, which consisted of Olympus Optical Company, Ltd. and its subsidiaries and affiliates, manufactured and sold opto-electronic equipment and other related products. The firm's major product lines included cameras, video camcorders, microscopes, endoscopes, and clinical analyzers. Olympus also produced microcassette tape recorders, laser-optical pickup systems, and industrial lenses. Olympus was founded in 1919 as Takachiho Seisakusho, a producer of microscopes. The brand name Olympus was first used in 1921 and became the firm's name in 1949. The first Olympus camera was developed in 1936, and by 1990 Olympus was

This case was prepared by Professor Robin Cooper of the Peter F. Drucker School of Management at Claremont Graduate University. The assistance of Professor Regine Slagmulder of the University of Ghent and Ms. May Makuda, KPMG Peat Marwick, is gratefully acknowledged.

TABLE 15-1. OLYMPUS OPTICAL 1995 FINANCIAL RESULTS

	Millions of yen			Thousands of U.S. dollars
	1995	1994	1993	1995
Net sales	252,097	239,551	267,718	2,801,078
Net income	3,101	556	3,805	34,456
Net income per share:				
Total assets	442,367	434,704	439,716	4,915,189
Working capital	205,256	202,070	164,712	2,280,622
Shareholders' investment	182,418	183,039	145,775	2,026,287

Notes:
1. Net income per share is shown in yen and U.S. dollars.
2. For the reader's convenience, U.S. dollar amounts were translated from yen at the rate of ¥90 = $1.
3. Fully diluted net income per share assuming full dilution is not presented because it is not significant.
4. The above figures were based on accounting principles generally accepted in Japan.

the world's fourth-largest camera manufacturer. Consolidated net income for 1990 was ¥8 billion, and total revenues were ¥219 billion.

Olympus had six divisions plus a headquarters facility. Four divisions—consumer products, scientific equipment, endoscopes, and diagnostics—were responsible for generating revenues (Table 15-1 shows 1995 financial results). The other two divisions were responsible for corporate research and production engineering, respectively. Headquarters was responsible for corporate planning, general affairs, personnel, and accounting and finance.

The consumer products division manufactured and sold 35mm cameras, video camcorders, and microcassette tape recorders. In 1995, the division employed 3,900 people (29% of the total Olympus work force) and generated revenues of ¥73 billion (29% of group revenues). Cameras were by far the firm's most important consumer product, accounting for ¥62.8 billion in revenues. Cameras were sold worldwide, with approximately 70% sold outside of Japan.

The consumer products division consisted of six departments: division planning, quality assurance, marketing, product development, production, and overseas manufacturing. Responsibility for the division's production facilities was centered at the Tatsuno plant, which opened in 1981 and was the firm's main camera production facility. Tatsuno was responsible for trial production of experimental products, introductory production of new products, and, to a limited degree, camera and lens production. Five other domestic manufacturing facilities reported to Tatsuno. These facilities were all located in Japan and were responsible for plastic molded parts, lenses, camera assembly, and die casting. Overseas production facilities located in Hong Kong and China reported to the overseas manufacturing department.

The 35mm Camera Market

Five Japanese firms dominated the world's 35mm camera market: Asahi Pentax, Canon, Minolta, Nikon, and Olympus. Canon and Minolta were the largest of the five firms, each with approximately 17% of the market compared to Olympus' 10%. There were two major types of 35mm cameras: single lens reflex (SLR) and lens shutters (LS) or compact cameras. SLR cameras, first introduced in 1959, used a single optical path to form the images for both the film and the viewfinder, allowing the photographer to see exactly what a picture would look like before it was taken. This ability allowed SLR cameras to take advantage of interchangeable lenses. Because of this feature SLR cameras rapidly gained a dominant share of the professional photographic market. As their price fell, they also came to dominate the high-end amateur market.

The low-end amateur 35mm market continued to be dominated by cameras with two separate optical paths. This market was divided into two segments. One segment contained very inexpensive cameras produced primarily by film manufacturers. The economics of this segment were driven predominantly by film, not camera, sales. The cameras in this segment primarily used the disc or 110mm film formats, though film producers had started to sell 35mm cameras that included new single-use versions. The other segment consisted of 35mm cameras that were less expensive than SLR cameras. This segment had undergone a dramatic change in the 1980s with the introduction of compact cameras.

Compact cameras, as suggested by their name, were smaller than SLR cameras. The first compact camera, the "XA," was introduced by Olympus in 1978 when miniaturized electronic shutters allowed the size of non-SLR cameras to be significantly reduced. The size of SLR cameras could not be equivalently reduced because their single optical path required a retractable mirror. This mirror was positioned between the lens and the film when in the down position and reflected the image into the viewfinder. When the shutter was pressed, the mirror retracted up into the body of the camera, allowing the image to expose the film. The retractable mirror, which was approximately the same size as the image, required SLR cameras to remain relatively bulky. Cameras with two optical paths, however, did not require a retractable mirror and therefore could be reduced to quite small sizes. For example, the Olympus Stylus, which was ergonomically designed to fit the hand, was only 4.6" long by 2.5" wide by 1.5" deep and weighed 6.3 ounces.

The early compact cameras were relatively unsophisticated and posed little challenge to the SLR market. However, as advances in electronic control systems allowed auto-focusing and automatic exposure features to be added at relatively low prices, the compact camera began to be viewed as a serious alternative to SLR cameras. The introduction of zoom auto-focus compact cameras in the mid-1980s removed the last major advantage of SLR cameras, that is, variable focal length lenses. Sales of SLR cameras plummeted.

The shift in consumer preference to compact cameras adversely affected Olympus in particular, because the firm historically had relied heavily on SLR sales and had failed to develop a leadership position in the compact camera arena. In the mid-1980s, Olympus' camera business began to lose money and by 1987 its losses were considerable. Top management ascribed these losses to a number of internal and external causes. The major internal causes were poor product planning, a lack of "hit" products, and some quality problems. While Olympus' overall quality levels were above average for the industry, certain products that relied on completely new technologies had rather high defect rates. These quality problems had caused Olympus' reputation to suffer. Externally, two factors were identified as primary contributors to losses: the appreciation of the yen from over 200 to the dollar in 1985 to around 130 in 1990 and an extended low-growth period for the industry that had caused prices (hence profits) to drop.

STRATEGIC CHANGE AT OLYMPUS

In 1987, Olympus' top management reacted to the losses by introducing an ambitious three-year program to "reconstruct" the camera business. At the core of this program were three objectives: first, to recapture lost market share by introducing new products; second, to dramatically improve product quality; and third, to reduce production costs via an aggressive set of cost-reduction programs.

Recapturing Market Share

To recapture market share, Olympus developed a new strategy of rapidly introducing, producing, and marketing new 35mm SLR and compact cameras. The firm's strategy in the SLR market was to differentiate its products from competitors' by the innovative use of technology. For compact cameras, Olympus' strategy was to develop a full line of low-cost cameras with particular emphasis on zoom lens models. Rapid introduction was considered important because it would allow the firm to react in a timely fashion to changes in the competitive environment. One of the key elements in improving the firm's ability to react rapidly was a plan to reduce to 18 months the time required to bring new compact cameras to market. The equivalent benchmark when the OM10 SLR camera was developed in 1980 was 10 years.

New products were introduced via the firm's extensive product planning process. At the heart of this process was the product plan, which identified the mix of cameras that the firm expected to sell over the next five years. The information required to develop this plan came from six sources: Olympus' corporate plan, a technology review, an analysis of the general business environment, quantitative information about camera sales, qualitative information about consumer trends, and an analysis of the competitive environment. As part of the three-year reconstruction program, the information collected to support the product plan was extended considerably from pre-1987 levels. In particular, the amount of qualitative data captured was increased. To ensure that all this new information was appropriately incorporated into the product plan, more extensive reviews of the plan were introduced.

The *corporate plan*, which was developed by Olympus' senior management, identified the future mix of business by major product line, the desired profitability of the corporation and each division, and the role of each major product line in establishing the overall image of the firm. It provided division management with a charter by which to operate.

The *technology review* had two sections. The first consisted of a survey of how current and future technological developments were likely to affect the camera business. For example, digital image processing was reaching the stage where electronic still cameras were rapidly becoming both technically and economically feasible replacements for conventional cameras that relied on chemical film for image capture. Olympus was in the forefront of electronic still image capture and in 1990 had introduced its first electronic camera. The second part of the review sought to determine whether Olympus had developed any proprietary technology that could be used for competitive advantage. For instance, Olympus had developed an advanced electronic shutter unit that combined auto-focus control and the lens system, which allowed the size of the camera to be smaller. This shutter unit allowed the firm to develop "small in size" as a distinctive feature of its cameras.

The *analysis of the general business environment* consisted of estimates of how changes in the environment would affect camera sales and the profitability of the business. Factors included foreign exchange rates, how cameras were sold, and the role of other consumer products. How cameras were sold was especially critical, because during the 1980s the percentage of the firm's cameras sold via specialty stores had decreased steadily from 70% to 40%. This change in retail distribution demographics had reduced the average wholesale prices of cameras because the bulk of cameras was now sold through discount houses and mass merchandisers, where profit margins were lower. The role of other consumer products was important because some of them competed for the same segment of the consumer's disposable income. For example, consumer research had shown that many consumers were trying to choose between buying a compact disc player or a compact camera. Therefore, Olympus viewed compact disc players as competitive products.

Quantitative information about the world's 35mm camera market was collected from three primary sources. The first was export and domestic market statistics for cameras published by Japan's Ministry of International Trade and Industry. These statistics included the number of units and dollar sales for each type of camera (e.g., zoom, SLR, and compact) for the entire Japanese camera industry. The second, published by the Japan Camera Industry Association, was statistics on camera industry shipments, which captured the number of units and dollar value of each type of camera shipped from the manufacturers to each major overseas market (e.g., the United States and Europe). The third source consisted of third-party surveys, commissioned by Olympus, of retail sales by type of camera in each major market.

Olympus collected *qualitative information* from seven major sources. First, the company collected questionnaires from recent purchasers of Olympus cameras. These questionnaires, included with every camera sold, captured information about the purchaser's age, income range, lifestyle demographics, and the other cameras the consumer considered before making the purchase. Second, group interviews were conducted by survey firms two to three times a year in each of the major markets to spot changes in consumer preferences for cameras. Third, surveys were conducted in Roppongi, the trendy fashion center of Tokyo; historically, these interviews had proven to be good predictors of future changes in the lifestyle of the Japanese population as a whole.

Fourth, professional photographers were interviewed to provide insights into both the leading edge of camera design and ways to improve the ease-of-use of compact cameras. Fifth, the Olympus sales force interviewed camera dealers. In addition, Olympus helped pay the salaries of "special salespeople" who worked behind the counters at very large camera stores. These individuals supplied Olympus with feedback about how their cameras were being received by consumers compared to competitive offerings. Sixth, members of the product planning staff would spend some part of the year behind the counter selling cameras, thus becoming familiar with the reactions of both consumers and dealers. Finally, members of the planning staff would attend industry fairs and conventions to obtain additional feedback on industry trends.

The *competitive analysis* was based on any information Olympus could gather about its competitors' current and future product plans. Sources of competitive information included press and competitor announcements, patent filings, and articles in patent publications. This information was used to predict what types of products competitors would introduce in the short and long terms and what their marketing plans were.

The information collected from all these sources was integrated into the preliminary product plan. This plan was the responsibility of a manager in the product planning section. Olympus differed from most other Japanese camera companies in the way it developed its product plan. First, the product planning function was part of sales and marketing, not research and development, as it had been prior to 1987. Second, the purpose of the product plan review was to balance the demands of a consumer-oriented market with the realities of research and development and production. Third, the firm had a stated objective of trying to design global products. Twice a year, the persons in the firm responsible for worldwide marketing met with the product planners to ensure that proposed products could be sold successfully in all the world's major markets.

Once the preliminary plan was completed, it was subjected to an exhaustive review to ensure its practicality. The review covered issues such as the expected sales volume and profit for each camera model and the load such sales would place on the division's production and research and development resources. A team composed initially of research and development and product planning personnel conducted the review. Subsequently, as the product plan approached acceptance, production personnel were added to the review team. Once the review was completed, a general meeting was held to formally accept the plan. This meeting was attended by division management, by the heads of the marketing, research and development, and production functions, and by the managers of the product planning section. If the product plan was accepted at this meeting, it was then implemented.

Improving Product Quality

Olympus' quality improvement program focused on two areas: the introduction of new products and the manufacturing process in general. The aim of the quality improvement program was to enable the firm to produce the highest-quality products in the industry. Olympus cameras were historically above average in quality, but management felt that it was important to be the best. Highest quality was considered important because it would help the firm recapture its lost market share by improving the reliability of the firm's products from the customer's perspective. In addition, improved product quality was expected to reduce production costs through decreased disruptions to the production flow.

Reducing Production Costs

To bring its high production costs into line, Olympus developed an aggressive cost-reduction program that focused on five objectives: to design products that could be manufactured at low cost, to reduce unnecessary expenditures, to improve production engineering, to adopt innovative manufacturing processes, and to shift a significant percentage of production overseas.

Designing High-Quality Products at Low Cost

At the heart of the program to design low-cost products was the firm's target costing system. The first step in setting target costs was to identify the price point at which a new camera model would sell. For most new products, the price point was already established. For example, in 1995 the simplest compact cameras were sold in the United States at the $80 price point, down from $100 in 1991. The actual selling prices for a given camera varied depending on the distribution channel (e.g., mass merchandiser versus specialty store). Thus, cameras at the $80 price point would sell for between approximately $70 and $100. The appropriate price point for a camera was determined by its distinctive feature (e.g., it might be magnification capability of the camera's zoom lens or the camera's small size). The relationship between distinctive features and price points was determined from the competitive analysis and technology review used in the development of the product plan. The product plan thus described cameras only in terms of their distinctive features. Other features were added as the camera design neared completion.

The price point at which a camera with given functionality was sold tended to decrease over time with improvements in technology. Price points were typically held constant for as long as possible by adding functionality to the cameras offered. Typically, a given type of camera would be introduced at one price point, stay at that price point for several years but with increasing functionality, and then as the functionality of the next higher price point was reached, drop to the next lower price point. The natural outcome of this process was to generate new price points at the low end. For example, the price point for the simplest compact camera was $150 in 1987 and $80 in 1995. At the high end, technology also generated new price points. As the functional gap between the capabilities of compact and SLR cameras closed, it became possible to introduce compact cameras at higher prices. For example, Olympus created a new price point of $300 when it introduced the first compact camera with 3X zoom capability in 1988.

The growing number of price points required camera manufacturers to expand their product offerings to maintain a full line. The decision to be full-line producers was based on two strongly held beliefs: that Japanese consumers trade up over time and that only by offering a full line could a firm obtain a balanced position in the entire market. A firm trying to compete in only the low end of the market would not have access to the high-end technology that would rapidly come to define the low-end market, and a firm selling only at the high end would not have the loyalty of consumers who were trading up.

The proliferation of products due to the increase in the number of price points was further aggravated by Olympus' decision to introduce multiple models for some price points. This change in strategy was prompted by the observation that market share associated with some price points was considerably larger than others. For high-volume price points, it was possible to identify different clusters of consumer preferences and profitably produce and market cameras designed specifically for those clusters. Under the new strategy, the number of models introduced at each price point was roughly proportional to the size of the market. Thus, the expected market share of each camera model offered was approximately the same unless it was designed to satisfy a low-volume strategic price point.

Once the price point of a new camera was identified, the free on board (FOB) price was calculated by subtracting the appropriate margin of the dealers and the U.S. subsidiary plus any import costs, such as freight and import duty. Target costs were established by subtracting the product's target margin from its FOB price. The product's target cost ratio was calculated by dividing the target cost by the FOB price. Every six months, the divisional manager set guidelines for acceptable cost ratios. These guidelines were developed in tandem with the division's six-month profit plans. In 1996, the divisional manager had identified the acceptable cost ratios as 85% for Tatsuno manufactured products and 60% for products manufactured overseas.

The target ratio for a given camera was set based on the historical cost ratios of similar cameras, the anticipated relative strength of competitive products, and the overall market conditions anticipated when the product was launched. Once the target cost ratio was established, it was converted into yen by multiplying it by the target FOB price. This yen-denominated target cost was used in all future comparisons with the estimated cost of production to ensure achievement of the target cost.

As part of the program to design low-cost products, target costs were set assuming aggressive cost reduction and high quality levels. A target cost system existed prior to 1987, but it was not considered effective. As part of the three-year program to reduce costs, the target cost system was improved and more attention was paid to achieving the targets. Aggressive cost reduction was achieved by applying three rationalization objectives. First, the number of parts in each unit was targeted for reduction. For example, the shutter unit for one class of compact camera was reduced from 105 to 56 pieces, a 47% reduction that led to a 58% decrease in production costs. Second, expensive, labor-intensive, and mechanical adjustment processes were eliminated wherever possible. Finally, metal and glass components were replaced with cheaper plastic ones. For instance, replacing metal components that required milling in an SLR body with plastic ones that could be molded reduced the SLR body costs by 28%. Similarly, replacing three of the glass elements with plastic ones in an eight-element compact camera lens reduced the lens cost by 29%.

During the design phase, the anticipated cost ratio of new products was monitored on a frequent basis, typically two to three times before launch. The FOB price of a new product was sensitive to both market conditions and fluctuations in foreign exchange rates. Olympus sold 70% of its cameras overseas, and the FOB price of a product was the weighted average yen price. Since the FOB price for cameras sold overseas was designated in the appropriate foreign currency, fluctuations in the exchange rates caused the FOB price to change when measured in yen.

If the FOB priced changed sufficiently during the design phase to cause the anticipated cost ratio for the camera to fall outside the acceptable range by about 10%, then the target cost of the camera was reviewed and usually revised to bring the anticipated cost ratio back into the acceptable range. If the FOB price was falling, the result was a lower target cost that was harder to achieve. If it was rising, the result was higher profits, which were used to increase promotions and advertising fees as well as reduce prices to overseas subsidiaries.

The target cost was based on the price point for the distinctive feature of the camera. Research and development was responsible for identifying the other features of the camera (e.g., the type of flash and shutter units). Feature identification was an iterative process in which the cost of each new design was estimated and compared to the product's target cost. Production engineering developed estimated costs of production in collaboration with production. Research and development reviewed these estimates and revised them as deemed appropriate. Most revisions resulted in lower estimated costs. The research and development group identified additional ways to reduce the cost of the product either through minor product redesign or a more efficient production process.

Approximately 20% of the time, the estimated cost was equal to or less than the target cost, and the product design could be released for further analysis by the production group at Tatsuno. The other 80% of the time, further analysis was required by the research and development group. First, marketing was asked if the price point could be increased sufficiently so that the target cost was equal to the estimated cost. If the price could be increased, the product was released to the production group. If the market price could not be increased sufficiently, then the effect of reducing the functionality of the product was explored. Reducing the product's functionality decreased its estimated cost to produce. If these reductions were sufficient, the product was released to production.

If it was not possible to raise the price or reduce the production cost enough to reduce the estimated cost below the target cost, then a life-cycle profitability analysis was performed. In this analysis, the effect of potential cost reductions over the production life of the product was included in the financial analysis of the product's profitability. In 1990, Olympus expected to reduce production costs by about 35% across the production lifetime of its products. The product was released if these life-cycle savings were sufficient to make the product's overall profitability acceptable. If the estimated costs were still too high, even with these additional cost savings included, the product was abandoned unless some strategic reason for keeping the product could be identified. Such considerations typically focused on maintaining a full product line or creating a "flagship" product that demonstrated technological leadership.

Once a new product had passed the research and development design review it was released to Tatsuno production for evaluation. The Tatsuno design review consisted of evaluating the research and development design to determine where and how the new product would be produced. To make these decisions, a detailed production blueprint was developed. This blueprint identified both the technology required to produce the camera and the components it contained. Using this blueprint and cost estimates from suppliers and subsidiary plants, the production cost of the product was reestimated. If this cost was less than or equal to the target cost, the product was submitted to the division manager for approval for release to production.

If the estimated production cost was too high, then the design was subjected to additional analysis. Frequently, relatively minor changes in the product's design were all that were required to reduce the cost estimate to the target cost level. As long as these changes did not change the product's price point, then the functionality was changed and the product was submitted for approval. If the design changes would change the price point, the product was returned to the research and development group for redesign.

The estimated production cost used in the evaluation of the product was the expected cost of production three months after it went into production. The initial cost of production was higher than this target cost due to the work force's lack of experience with producing the new camera. As the work force gained experience, production costs would fall below target costs. Thus, the cost system would report negative variances for the first three months. In subsequent months the variances were expected to be positive. After the product was in production for six months, the target cost was changed to reflect any expected savings in the next six months due to the firm's cost-reduction programs.

Reducing Unnecessary Expenditures

The program to reduce unnecessary expenditures had four components.

- It analyzed fixed expenses and curtailed any unnecessary expenditures.
- It analyzed and improved the procedures surrounding new product launching to reduce launch costs.
- It lowered the cost of purchased parts by implementing strict controls to ensure that target costs were met, widening the sources of procurement to obtain lower costs, and identifying multiple suppliers for each component to create competitive pressures.
- It strengthened and integrated its existing cost-reduction programs.

Olympus relied on four programs to control and reduce costs. The first program focused on production costs, the second on the costs of defects, the third on capacity utilization costs, and the fourth on overhead expenses. The *production cost control and reduction program* focused primarily on removing material, labor, and some overhead costs from products; the division's profit plan identified cost-reduction targets for these costs for each product. These targets were considered challenging though achievable. The standards were set every six months and included the anticipated reductions that would be achieved in the next six months. Progress toward achieving these cost-reduction targets was monitored using variance analysis. Material price, work improvement, and "budgetary other" cost variances were computed weekly and accumulated monthly.

The material price variance was computed for each product by comparing the actual material cost to the standard material monthly target. This target was the average of the material costs for the previous six months adjusted for any anticipated changes in material costs in the upcoming month. The work improvement variances were the difference between the actual labor hours and the standard labor hour monthly target and between actual machine-hours and the standard machine-hour monthly target. Their target was calculated by assuming that labor cost reductions would occur evenly over time. To these linear cost reductions were added any specific reductions due to planned changes in the production process. The actual "budgetary other" costs, which included general expenses of the factory, were compared to budgeted costs to determine the other budgetary variance.

The second cost control and reduction program focused on the *costs of defective production*. To give these costs high visibility, they were not included in the standard costs and hence were not covered by the production cost control and reduction program. The cost of defects program consisted of setting cost of defects targets for each production group every six months. Groups were responsible for segments of the production process (at Tatsuno there were 10 groups).

Cost-reduction targets were identified for each product the group produced. Division management negotiated with the group leaders to set cost-reduction targets for each product the group produced. The group leaders recommended their cost-reduction targets, then divisional management reviewed these recommendations. If the overall reductions were sufficient to achieve the division's cost-reduction objectives, divisional management accepted the targets. If the overall savings were insufficient, the targets were renegotiated until the savings were acceptable.

The team leader and foreman in each group met daily to discuss their progress at achieving their reduction targets. Group and team leaders held weekly meetings to report on progress. If a group did not meet its weekly objectives, the group leader was expected to explain why the group had failed and what corrective actions would be taken. A request to engineering for assistance might be included in these actions. Occasionally, if a group consistently failed to meet its objectives, management would send in engineering—a serious blow to the group's reputation.

The third program focused on managing the costs associated with *capacity utilization*. The division's long-range management plan included estimates on the amount of overtime, actual working hours, operation days, and attendance rates. These estimates and the expected workload for each cost center were combined to give a capacity utilization cost budget for each center. This budget, which consisted of overall attendance rates and direct labor hours by cost center, was set every six months and updated each month. The updated monthly budget was used to compute a daily variance, which was reported to management weekly and accumulated monthly. The variance captured the over- or under-utilization of direct labor capacity at the standard distribution rate of processing costs.

The final program focused on *overhead expenses*. These expenses included items such as the personnel expenses of support and administration, depreciation of factory buildings, and computer costs. The long-term management plan contained targeted levels for these expenses. Monthly budgets for these expenses were prepared taking into account the production volume for the six-month period, the introduction of new products, and any planned cost-reduction actions by the groups. Division management approved the resulting budget after any necessary adjustments were made. Each month, the budget was compared to actual and multiple cost center variances. Costs subjected to separate variance analysis included machine repair costs, machine maintenance costs, expenses of repair and maintenance personnel, and miscellaneous expenses. These variances were computed monthly because management felt that these expenses could not be controlled in a shorter time frame.

The four cost control and reduction programs each generated variances, which were combined in a monthly cost report. This report provided division management with important insights into the success of the cost control and reduction programs.

Improving Production Engineering

Olympus achieved the desired improvements to production engineering through a three-phase approach. This approach shortened production lead times by decreasing batch sizes. In the production area, for example, batches were halved and moved to a zero inventory system. Improving communications between sales and manufacturing reduced introduction times for new products and production lead times in general. For example, the MRP system was used to check inventory levels twice a day as opposed to once a week. Finally, office automation improved the level of general administrative support provided to both marketing and sales.

Adopting Innovative Manufacturing Processes

The program to introduce innovative production technologies focused on increasing the level of automation in manufacturing, particularly in the assembly, lens production, electronics parts mounting, and molding processes. In all these processes, the level of automation was significantly increased. For example, in assembly four major processes were automated in the three-year period after the new strategy began: the assembly of the film winding and shutter units, the adjustment and inspection processes for the focusing unit, the alignment and related inspection processes, and the transportation system for assembled parts.

Similarly, the molding, lens processing, and IC mounting stages of production underwent complex changes. All told, the program initiated some 23 different automation projects.

Shifting to Overseas Production

The cost reductions that the aggressive application of target costing and production cost reduction achieved were further augmented by shifting some of the manufacturing processes to lower-cost areas of the world. Olympus was the last of the camera firms to open such overseas facilities. Other manufacturers had opened such facilities in the late 1970s and early 1980s. Cost analyses at Olympus had indicated that the potential savings from shifting production offshore was about 15%. In 1988, the firm opened production facilities in Taiwan, Hong Kong, and Korea, and in China in 1989. The firm anticipated offshore production to reach ¥10 billion by 1991 and to expand rapidly thereafter.

The New Cost-Reduction Effort

The 1987 program to reconstruct Olympus' camera business achieved most of its objectives. The program to introduce new products was relatively successful at recapturing lost market share. The firm increased camera sales volume by almost 70% to ¥50 billion from ¥30 billion and almost doubled its market share for compact cameras. Unfortunately, the program was not as successful for the SLR product line. The firm continued to lose market share from 1987 through 1990, but with the introduction of a completely new camera, the IS-1, the firm hoped to turn the situation around.

The combined results of the cost-reduction program were impressive. By the end of 1990 every measure of productivity at Tatsuno had improved. For example, overall production had increased by 50%, the production cost ratio had fallen by 20%, the production value per employee had risen 70%, and gross added value per person had increased over 125%. Simultaneously, the work-in-process inventory had not increased despite the higher activity level and the fact that lead time had almost halved.

Despite the success of the 1987 plan, top management at Olympus determined that additional cost reductions would be necessary in the coming years. In particular, they were worried by three trends that together would place significant pressure on the firm's profitability. These trends were an increased proliferation in products required to satisfy consumer demand in the domestic market, an additional shortening of the product life cycle to less than a year, and reduced selling prices. The decision to introduce a new program was driven in part by the observation that the savings from the 1987 plan had gone down in recent months.

At the heart of the new plan were two important concepts. The first was innovations in technology, and the second was functional group management. Innovations in technology consisted of applying new production technology—primarily automation—to all stages of production. Separate automation projects were initiated for camera assembly, lens processing, molding, and electrical components. The most ambitious of these projects was a fully automated robotic assembly line designed to assemble cameras. This line was undergoing evaluation at Tatsuno before being released to other assembly facilities.

Functional group management consisted of dividing the production process into a number of autonomous groups. Ten such groups were identified at the Tatsuno plant. These groups were given full management responsibility for their area of responsibility or cost center and were expected to manage it as if it were a separate company. Thus each group would effectively become a separate profit center. Top management felt that holding the groups responsible for their profitability would promote greater pressure to reduce costs and hence increase profitability than would any conventional cost-reduction program. By 1990, senior management had yet to operationalize the function group management concept but believed that it was going to play a critical role in the firm's future.

APPENDIX

The Evolution of the Cost System

From 1970 to 1990, the firm's cost system had undergone three major changes. Prior to 1976, there was only one overhead rate at the Tatsuno plant. The system directly traced some material costs to products, but all other costs were allocated. These allocated costs were divided into two categories: processing and overhead. Processing costs included the indirect material, direct and indirect labor, and direct expenses of the production process. Direct labor was allocated because the direct labor wage rates varied by individual, and it was considered too expensive to assign the cost directly to products. The overhead costs contained the indirect material, indirect labor, and indirect expenses associated with support and administration. The processing overhead costs were combined and divided by the number of direct labor hours to give an average allocation rate. The reported cost of a product was given by the sum of the direct material charge and the direct labor hours that the product consumed multiplied by the allocation rate. Such a simple system was considered adequate because there were only small differences in the cost structure of the products. In addition, the level of automation was small, as was depreciation. The stated objective of this system was to differentiate material cost from other expenses and provide mechanisms for total cost reduction.

In 1977, the cost system was updated. The overhead costs were split into two categories: procurement costs and other costs. Procurement costs were those costs associated with obtaining raw material and purchased parts. They included the personal expenses of the procurement section, transportation charges, car fares, and other miscellaneous expenses. A single allocation rate was determined for processing costs and some of the other costs. A separate rate was determined for procurement costs and the allocated expenses of the administration and production technical sections. The costs of these two sections were allocated to the production and procurement sections based on head count. The procurement costs were allocated to products based on the sum of the direct material charge plus the allocated processing costs.

The primary purpose of this system was to draw attention to the procurement costs, which had grown substantially over time. This increase was due both to an increase in production capacity, which was accompanied by a corresponding increase in the volume of procured parts, and by an increase in the ratio of procured to internally manufactured parts. The other important change in the system was its focus on the cost of quality. The cost of defects was isolated from the standard costs to give it more visibility. Separate variances were computed for standard production and defects.

In 1983, the cost system was again updated. The general structure was maintained, but now multiple allocation rates were computed for processing costs. The production process was split into 10 different cost centers and different overhead rates were computed for each center. Examples of the cost centers included camera final assembly, electronic flexible board assembly, lens processing, and lens assembly. In addition, the firm had begun to enter into OEM contracts with other firms that would produce components for Olympus. The support and administration costs for the OEM production were significantly different from Tatsuno production. To capture this difference, the two overhead cost allocation rates were computed, one for general suppliers and the other for OEM suppliers. These two rates replaced the single procurement rate computed in the prior system. The treatment of other costs as partially related to processing and partially related to procurement was suspended, and all other costs were allocated as part of the support and administration costs. The primary purpose of this system was to provide improved control over production and support and administration costs.

CHAPTER 16

SONY CORPORATION: THE WALKMAN LINE

INTRODUCTION

Sony Corporation, one of the world's largest electronics companies, started in 1945 as the Tokyo Tsushin Kenkyujo, or Tokyo Telecommunications Research Institute. To survive in its earlier years, the firm generated revenues by repairing broken radios and manufacturing short-wave converters that allowed medium-wave radios to receive short-wave transmissions. By 1946, the firm was strong enough to incorporate and became Tokyo Tsushin Kogyo (Totsuko), or Tokyo Telecommunications

This case was prepared by Professor Robin Cooper of the Peter F. Drucker School of Management at Claremont Graduate University. The assistance of Professor Regine Slagmulder of the University of Ghent is gratefully acknowledged.

Copyright © 1994 by the President and Fellows of Harvard College. Harvard Business School case 195-076.

Engineering Corporation. The new company, capitalized with only ¥190,000, had no machinery and little scientific equipment. Its stated mission was to research and manufacture telecommunications equipment and voltmeters.

The firm's first few products were developed solely to ensure a steady stream of cash to help pay for salaries and other necessities. There was no underlying strategy to the development of these products, which included a rice cooker, an electric cushion, a power megaphone, and assorted electronic components and products primarily for government use.

The company's first really successful product was a magnetic tape recorder (Japan's first). It was brought to market in 1950 after a year of intensive research and development. Initially, the firm had problems selling the product because it was large, bulky, and not very reliable. Subsequent models were smaller and lighter. The reliability problem was temporarily solved by the creation of a 12-person service department that visited users and kept the tape recorders running by performing preventive maintenance. As the design of the product improved, both sales and the firm's reputation grew.

In 1953, Totsuko signed an agreement with Western Electric for a nonexclusive patent license for use of the newly developed transistor. Western Electric advised the firm to make products such as hearing aids, where small size, low volume, and high price were compatible. Instead, Totsuko management sought to be the first company in the world to market a transistor radio, which was a high-volume, low-price product. However, the first company in the world to market a transistor radio was Regency, an American company. Totsuko brought its first transistor radio to market approximately a year later and rapidly developed a better reputation than Regency. It soon dominated the world market.

Totsuko grew rapidly after introducing the transistor radio. The firm was listed on the over-the-counter market of the Tokyo Stock Exchange in 1955 and on the Exchange itself in 1958. Also in 1958, the firm changed its name to Sony, a brand name it had used for several years for some of its domestic products including magnetic tape (Soni-tape) and the TR-55 transistor radio. Part of the motivation for the name change was that unlike Totsuko, Sony could be easily pronounced in English.

The firm continued both its rapid growth and fast pace of product innovation. It became a truly international firm in 1960 with the opening of Sony Corporation of America and Sony Overseas, S.A., in Switzerland. Sony (UK) Limited followed in 1968, and Sony GmbH was formed in 1970. Overseas production began in 1972 with the opening of the San Diego facility in the United States and in the U.K. in 1974 with the opening of the Bridgend, Wales, facility.

THE WALKMAN PRODUCT LINE

The first Sony Walkman, the TPS-L2, was introduced to the Japanese market on July 1, 1979. The product, a fusion of a small playback-only tape cassette player and lightweight headphones, was an immediate success. It created a completely new type of product, the personal audio system. In the period from 1979 to 1994 (the 15th anniversary of the Walkman), over 250 million personal audio systems were sold. Sony alone accounted for more than 120 million units. In 1996 the cassette tape still remained the dominant media, by far, for personal audio systems. However, Sony anticipated that the 64mm recordable magneto-optical disc format called the Mini Disc, which it had introduced in November 1992, would eventually become the dominant media.

The project to develop the Walkman began in the fall of 1978, when Masaru Ibuka, honorary chairman of Sony, walked into the office of Akio Morita, chairman and chief executive officer of Sony, carrying a Sony cassette tape recorder and a set of headphones. Ibuka had been using the tape recorder and headphones to listen to music privately. He complained to Morita that the combination was too bulky and heavy for easy use, especially on planes.

Morita and Ibuka discussed ways that the marriage between a cassette player and headphones could be made more practical. The two went to the engineers of what is now known as the general audio business group and asked them to redesign the cassette tape recorder for stereo playback. The general audio group's engineers modified an existing tape recorder by eliminating the recording functions and speakers, leaving only the desired stereo playback function. Although the engineers were not sure a product without such fundamental functions as recording and speakers would sell, the resulting design provided such high-quality sound that Sony decided to take a chance and commercialize the product.

Fortuitously, another research team was already working on a set of lightweight headphones. This team had managed to produce a prototype set of headphones that weighed only 50 grams, which was about half the weight of the lightest set then on the market. The prototype Walkman, a merger of the new playback-only machine and the light headphones, was produced in under a year at Ibuka's and Morita's request. The reaction of virtually everyone who tried the prototype was positive. When Morita himself tried the product, he said, "If it doesn't sell well, I'll resign as chairman," and so the firm became committed to launching the new product.

The product was named only a month and a half before it was launched. The name of the product in the domestic market, Walkman, was chosen to capture and communicate the unique character of the product. The name was a combination of the ability to walk with the product and the Pressman, the lightweight recorder and player that Sony had introduced the previous year (from which the technology that allowed the Walkman to be produced in such a short time was derived).

Once the new product's name was chosen and preproduction prototypes were available, the advertising department showed the product to a number of its Tokyo dealers for their reactions. This was the first time that any outside reaction to the new product had been sought. Eight out of ten dealers surveyed reacted negatively. However, the younger sales staff and the dealers' children reacted favorably.

For the next few weeks, Sony had some popular Japanese singers try the Walkman. The entertainers took immediately to the product and, fortuitously, were photographed using it; these photographs appeared in the popular entertainment and weekly magazines. The photographs, combined with the large number of young Sony employees who were using prerelease versions of the product on their train rides home and around fashionable areas of Tokyo such as Ginza, created high public awareness of the product even before it was launched.

Sales of the Walkman were slow for the first 20 days it was on the market. However, once sales started, they increased rapidly—within three months, the entire prelaunch inventory of 30,000 units was sold, and production could not keep up with demand. The demand never stopped, and the TPS-L2 sold over 1.5 million units in just two years.

The first competitive product was introduced by Aiwa more than 12 months after the introduction of the TPS-L2. This long delay in the emergence of a copycat product was unusual for the Japanese home electronics appliance industry, which was renowned for its ability to introduce such products within a few months of the original's appearance. The long delay was caused both by skepticism among Sony's competitors about the sales potential of Walkman-like products and by the need to develop designs for their own competitive products.

Sony introduced a second-generation product, the WM-2, in February 1981. The WM-2 was the first product to use the Walkman name worldwide. While the TPS-L2 was called the Walkman in Japan, it was called the Soundabout in the United States, the Stowaway in the U.K., and the Freestyle in Australia. In retrospect, Sony decided that a single worldwide name was preferential in order to create maximum brand image.

The WM-2 established both the personal audio system and the dominance of the Walkman line throughout the world. The WM-2 was launched simultaneously in Japan, Europe, and the United States via a worldwide sales campaign. Reaction to the product in Europe and the United States was overwhelmingly positive. The WM-2 rapidly became the best-selling Walkman in history, selling over 2.5 million units before being replaced by the WM-20.

The success of the WM-2 reflected the heavy emphasis that Sony engineers placed on compactness, high performance, and superior sound quality. Not only was the WM-2 smaller and lighter than the product it replaced, but it also produced higher-quality sound. For example, the WM-2's player weighed only 250 grams and its headphones weighed 28 grams; in comparison, the TPS-L2's player weighed 435 grams and its headphones weighed 45 grams. When introduced, the WM-2 not only beat the full-scale market entry of other manufacturers but also set a technological standard that was hard to match, let alone exceed.

The introduction of the WM-2 signaled the beginning of the proliferation of the Walkman line. By the end of 1981, Sony manufactured and sold five Walkman models in three different product categories (playback only, record and playback, and playback and tuner). The higher number of models than categories reflected multiple product variations in two of the categories. In 1982, the proliferation of the Walkman line continued. The first professional model, the WM-D6 (an upscale version of the WM-2), was introduced in February 1982. In October 1982, the WM-DD model was launched. This model produced even higher-quality sound than the WM-2 but still retained the same exterior size. The market enthusiastically received both the WM-D6 and WM-DD.

Throughout this period, Sony's competitors had been active and continued to intensify their challenge. Each firm was dedicated to developing the world's most compact player. Every model saw a reduction of 1mm to 2mm in the width of players. In October 1983, Sony introduced the WM-20, the world's most compact player. At the time, the WM-20 was a technological marvel. It was the same size as a tape cassette case; its body weighed only 180 grams and its headphones only 17 grams. It was half the size of its predecessor and used only one AA battery instead of two. Reducing the number of batteries by half meant that the mechanical components in the WM-20 had to deliver twice the performance of the equivalent components in the WM-2 despite being half the size.

THE DOMESTIC MARKET

The domestic market was dominated by high-end products designed to last several years despite heavy daily use. They were used primarily by young urban commuters to alleviate boredom on their long train journeys to work. In 1996, there were three major domestic competitors: Sony, the market leader, had a 45% market share; Matsushita, the parent of Panasonic, had a 30% market share; and Aiwa had 20%. Sony's market share had remained relatively constant for over five years. The bulk of the remaining 5% of the market was shared by Kenwood and several non-Japanese competitors. Three Japanese firms, Toshiba, Sanyo, and Sharp, had entered and then exited the market after being unable to achieve significant market penetration.

Sony's relentless pursuit of improvement was demonstrated by the introduction of the Stamina line. The extended play times for the Stamina models were achieved by radically simplifying the tape drive mechanisms from two capstans to one. The original Stamina could play for 7 hours on just its rechargeable battery, and 22 hours on a combination of its rechargeable and replaceable batteries. In 1994, the WM-EX1 (the 15th anniversary model) was launched. This Stamina version could last 12 hours on just its rechargeable battery or 36 hours using both its rechargeable and replaceable batteries. In 1996, the WM-EX2 was introduced. This Stamina could run for 50 hours using both its batteries.

Not all attempts to proliferate the Walkman line were successful. There were several notable failures, including the dual cassette model and the rewinding headphone model. The dual cassette model provided the user with the ability to record from one tape onto another. It was necessarily thicker than existing models. Consumers reacted negatively to this model, and very few were ever sold. Consumers had the same reaction to the automatically rewinding headphone model; while it had the added advantage of automatically retracting the headphone wire when the user was finished listening, it, too, was much thicker than other models. Apparently, the thickness of a Walkman was a critical attribute. The sports model, which was water-resistant, was not particularly successful in Japan but was very successful in the United States and Canada.

The typical Walkman model was sold for only about 12 months before a new version was introduced. Typically, each generation was similar to the one that preceded it unless new functionality was being introduced to improve the product. For example, in 1987 Walkman stereos that used wireless headphone technology were introduced, and in late 1992, the WM-EX 909, or Stamina, was introduced. These two improvements were designed to make the Walkman easier to use in crowded trains. The wireless model removed the need to have the headphones connected to the Walkman. The Stamina was designed to play for extended periods of time between either recharging its nickel-cadmium or nickel metal hydride rechargeable battery or replacing its one AA dry cell.

The only Walkman stereos that were not replaced every year were the professional models. They were viewed originally as strategic products and not expected to be profitable. They were introduced to demonstrate Sony's dedication to the production of high-fidelity sound. As it turned out, professional models had much longer product lives than their ordinary counterparts. For instance, the most successful professional model was introduced in 1982 and was still selling in 1995. The longer product lives and their higher prices allowed the professional models to be profitable despite their lower production volumes.

The range of models that Sony produced was designed to strike a balance between profitability and customer service. Too many models would cause production and distribution costs to be too high. Too few models, and too many sales would be lost to competitive products. The general stability of the number of Walkman stereos in the product line indicated Sony's belief that the number of models it offered was optimal. The exact number of models produced for the domestic market had been relatively stable since 1985. At that time, there were 18 models in five categories (playback only, playback and record, playback and tuner, professional playback only, and professional playback and record). In 1992, seven years later, Sony still manufactured 18 models for the Japanese market: 5 playback (3 conventional and 2 water-resistant), 3 with built-in tuners, 4 professional, 2 wireless, and 4 recording models.

In 1993, the slowdown in worldwide sales of consumer electronic products created pressure on Sony to reduce its overall product offering. In particular, Morita, Sony's chairman, had argued that Sony should reduce the number of its product variations as well as extend their life expectancies. The problem of reducing the number of Walkman models and extending their life cycles was exacerbated by the nature of the Walkman market in Japan. The primary consumer was the young urban commuter. Commuting was not common in rural areas, and the older commuter tended not to use headphones. The young urban commuter was highly fashion-conscious. Some 60%-70% of sales consisted of these young consumers replacing an existing product that they had purchased only two or three years earlier. These persons wanted the latest technology and would buy only new models. This tendency made it difficult to reduce product offerings or extend product lives without Sony losing competitive advantage.

In addition, there was some evidence that the Walkman line was too small. Market surveys indicated that 70% of consumers picked Sony as their first choice. Sony called this its "mind share." Sony viewed the difference between its 70% mind share and its 45% market share as a lost opportunity. Most of this lost opportunity was due to the lower prices of competitive products. Sony charged a premium for its products because of its strong reputation as the originator of the product and its technical leadership. The level of customer loyalty was not known because Sony did not maintain a database on repeat purchases.

Consequently, Walkman executives were convinced that dropping any one of the models would leave a distinct hole in the product lineup and would cost Sony market share. For example, in 1993, there were five playback models, each of which had a very specific role in establishing the market. The two cheapest models created the market for the most popular model. The other two playback models each had specific properties that satisfied a particular segment of the market: the water-resistant model was used by sports enthusiasts and the auto-repeat model was popular with foreign language students because it allowed them to continuously listen to language tapes.

The other categories of Walkman models were similarly serviced by a careful structuring of the models offered. The careful segmentation of the market showed in the relative sales volume of each model. With the exception of the professional models, all Walkman models sold in relatively high volumes. The highest-selling model accounted for only about 10% of the total nonprofessional sales and the lowest, for less than 1%.

The most obvious candidates for discontinuance were the professional models because of their low sales volumes, which were usually less than 2,000 units per month, and the emergence of new digital technologies that provided superior sound quality to their analog predecessors. By 1994, two digital products were available, the minidisc and the NT player. The minidisc used 64mm recordable discs and the NT, postage-stamp-sized magnetic tapes. In 1995, all but one of the professional players were phased out, and in 1996, the WM-D6, the original professional model, was finally withdrawn.

The withdrawal of the professional models reduced the number of Walkman stereos on sale in the domestic market. There were now seven playback, each designed to provide a different range of performance in terms of battery life, headphones, auto reverse speed, design, and price, three with built-in tuners, two wireless, and four recording models. The increase in the number of playback-only models was considered necessary to satisfy current market needs. The three mainstream playback-only models created the market for the most popular model, the WM-EX2. The other three playback models each had specific properties that satisfied a particular segment of the market: the WM-EJ95, a water-resistant model, was used by sports enthusiasts; the brightly colored WM-EQ2 was designed for the youth market; and the WM-EX911 had headphones that reeled into the body of the player.

THE OVERSEAS MARKET

The overseas market contained two types of competitors, the Japanese firms (dominated by Sony with its 40% share) and a large number of

Asian competitors, many of whom were relatively unknown. With its strong quality and reputation for innovation, Sony was able to charge premium prices in the overseas market but was still faced with significant downward price pressure.

There were two significant differences between the domestic and overseas markets. In Japan, the Walkman was used primarily to provide entertainment during the long commutes that characterized big city life in that country. In Tokyo, for example, it was not unusual for commuters to travel for more than three hours a day by public transport. To satisfy this use, size was critical—the smaller, the better. In contrast, overseas the Walkman was typically used either indoors or for sports activities. Consequently, the typical Walkman sold in the overseas markets was larger than those sold in Japan.

Another major difference between the two markets was the amount that the consumer was willing to pay for the product. In Japan, the minimum price that the consumer was willing to pay was about ¥15,000 and the most popular price was about ¥20,000. In the overseas markets there was no effective lower price. For example, in the United States the cheapest portable cassette player, which was produced in the Republic of China, sold for $9.99.

Such low-price products did not sell well in Japan. The price/demand curve in Japan did not follow the usual shape portrayed in economics textbooks (i.e., the lower the price, the higher the demand). In Japan, demand was virtually nonexistent below ¥15,000; above that price, demand increased until the price reached ¥20,000. After ¥20,000, demand fell with increasing price.

Some of Sony's competitors, particularly Aiwa, had tried to introduce low-price products into the Japanese market but were not successful. The Japanese consumer did not like "cheap" products and preferred not to buy the least expensive product available. Instead, the majority liked to buy the second or third model in the range, hence the success of the ¥20,000 model.

The demand curves for the European and U.S. markets more closely fit the classical model: the lower the price, the higher the volume. This difference in consumer behavior toward price defined the nature of competition in the two markets. For example, the most popular Walkman in the U.S. market listed for $26.95 and typically sold for under $25.00, or about one-eighth of the typical selling price of the most popular domestic model.

Although the domestic market could be served by 14 to 15 nonprofessional models, the overseas markets required 49. These models were spread across four categories: 12 playback-only, 21 built-in tuner and playback, 7 recording, and 9 sports models. The higher number of models reflected the need to design products for specific regions of the world. The number of products sold in overseas markets was also essentially stable. In 1992, for example, 50 overseas models were offered, only 1 more than in 1996.

THE TARGET COSTING SYSTEM

The target costing system was used to determine the allowable cost of a product before it was designed. The target cost was set in a two-stage procedure. First, the target price of each product was calculated by subtracting the dealer's and the wholesaler's margins from its price point plus 10%. Next, the target cost was calculated by subtracting the target margin for that product from its target price.

The suggested retail prices of the Walkman products were set so that after discounting they sold at the so-called magic price points. These were prices slightly below a round number; for example, the magic price point of Sony's most popular model in the domestic market was ¥19,800, just below ¥20,000. Experience had shown that pricing products at the magic price points significantly increased their sales volumes. To capture this effect, Sony's suggested retail prices for its products after appropriate discounting were at the magic price points. The selling price of Sony Walkman stereos was not affected by the selling price of competitive or alternative products. If a competitor dropped prices, Sony usually would respond by accelerating the launch of new products with enhanced functionality, rather than by matching prices. For alternative products, such as cameras and CD players, the Walkman appeared to be the benchmark product against which the other products' prices were set. To maintain this advantageous benchmark position, Sony set Walkman prices based on the consumers' perceived value of the product to maintain maximum customer satisfaction.

The target margin for each product was determined by an iterative process designed to ensure that overall the division achieved its long-term profit objective. The starting point was the group profit margin as identified by the group profit plan. This plan was developed by Sony's corporate planning department after negotiations with the planning department of the general audio group. Once the annual group profit target was set, the group was responsible for its own profitability.

The first-cut target cost for a new product was calculated by subtracting the group target profit margin from the product's target selling price. The resulting target cost was compared to the estimated cost of the new product. When the target cost was considered too high, then the target margin was allowed to decrease if another product with a sufficiently high target margin could be identified to offset the loss. When all the individual product decisions were completed, a simulation of overall group profitability was run to ensure that the group's target would be met. If the group target could not be met, then the process was repeated until acceptable target costs had been established that would enable the group to realize its profitability objectives.

If the planning process indicated that the group's target would not be met, then the individual decisions that had been made in the process of establishing the group's plan were reviewed. Three steps were considered when this occurred. First, the engineers looked for other ways to reduce the product's cost. Second, actions taken to improve group profitability included asking marketing to increase prices and reducing the functionality of low-margin products. Third, the general audio group and the marketing group would both accept reduced profitability if the product in question was considered critical to the corporation but could not achieve its target margin.

The way the target cost system operated at Sony reflected the relatively small capital investment required to produce a new product and the short product development cycle of about one year. Most new Walkman stereos were direct replacements for existing models with relatively little innovation, hence the low capital investment and short development cycles. The only exceptions were products that relied on new technologies, such as the wireless models, or on radical new engineering solutions, such as the Stamina. These products represented greater investments and typically had more extensive product development cycles.

Because capital investments were small, Sony could afford to experiment with different models; the failure of a single model did not adversely affect profitability. Short development cycles meant that when Sony was estimating the sales volume, selling price, and production cost of a new product it usually had information about the previous year's model to serve as a starting point for its estimates. Because the new product was usually designed to replace the previous year's product and was similar, the previous product's sales volumes and selling prices were good predictors of the success of the new model. Consequently, Sony product planners were relatively confident about most of their estimates. This confidence allowed them to make profit-margin trade-offs across products rather than requiring each product to achieve its target cost. They did not have absolute freedom in relaxing a product's target cost. As a matter of policy, Sony would not sell products at a loss and, under most conditions, would not sell them below the minimum profit margin established by the appropriate business group's manager. The only exceptions to this rule were strategic products, which Sony top management viewed as investments necessary to create or expand markets and which would pay off in the long run.

New Product Development

New product development for Walkman stereos was shaped by Sony's overall strategy of being a technology, not price, leader. Sony viewed itself as a technology pioneer with no equal and tended to compete on product functionality, not price. In markets like the Walkman, it tended to compete more with itself than other firms. It would aggressively try to make its own products obsolete by bringing out new models with enhanced functionality with the objective of maintaining the highest level of customer satisfaction possible. Based on the enhanced functionality of its products, their reliability and brand image, Sony Walkman stereos typically sold at a slight price premium over competitive products.

Sony was widely considered the technical leader in the Walkman market. It had consistently introduced the benchmark products ahead of its competitors. Given this leadership, Sony did not systematically tear down its competitors' products, only those that it found technologically interesting. Usually, there was very little Sony could learn from its competitors about the design of personal audio products.

To maintain its strategy of being a technology leader, Sony competed only in markets where products had yet to reach technological maturity. Such markets allowed technology to be used to create products with significantly enhanced functionality that would be well received by the customer. In such markets, Sony usually did not undertake extensive consumer analysis because it believed that customer requirements in such markets were either obvious or were unknown to the consumer. For example, the development of the Stamina was not based on extensive consumer analysis but on the general awareness that battery life was considered critical. This assumption proved to be correct, the revolutionary design of the Stamina caused it to become the industry standard against which competitive products were measured. In contrast, the original Walkman, again undertaken with no consumer analysis, created its own market: portable headphone stereos. Sony relied on its strength and experience in product planning rather than on consumer analysis to propel its success.

Sony's new product development process began with the construction of a handmade model. The purpose of this model was to ensure that components would fit into the case and function together effectively. The successful development of a handmade model was a major decision point in the development of a product because the next step was to develop the injection-molding dies. These dies were necessary for the mass production of the case and other plastic parts and required a high level of capital investment.

Once the product was approved, the parts list was developed and provisional orders were placed. The suppliers and subcontractors accepting these provisional orders provided more detailed information on the cost of the parts required by the new model. Sony prided itself on maintaining supportive relationships with its suppliers, many of whom were 100% owned. Typically, relationships with suppliers were maintained over an extensive period of time.

The estimated cost of the new product was continuously compared to its target cost. The target costing system thus acted as an important link in maintaining group profitability. It was designed to monitor the relationship between the selling price of a Sony Walkman and its manufacturing cost. If it appeared that the new product would fail to meet the firm's minimum profit margin requirement or would adversely affect the ability of the group to meet its profit target, then the product would be redesigned.

Sony took two approaches to product redesign. The first approach was simply to start again. This approach was adopted if the design problems were identified very early in the design process and could not be rectified by partial redesign. The second approach was an iterative process in which the costs were removed from the product by redesigning parts of it. This technique was used if the cost overrun was small or was discovered too late in the process for a complete redesign to be feasible.

The driving force behind which approach—complete redesign or partial redesign—was chosen was the timing of the product introduction. Sony believed that it was imperative to release products on a timely basis. Consequently, it did not allow product redesign to extend the launch date. The Walkman market was so competitive that failure to release a new model on a timely basis usually would result in considerable lost sales. Because the physical production facilities would still exist, the firm saw no benefit to missing a launch date. Consequently, it would launch a product even if its profitability was below the minimum level in order to meet deadlines.

PRODUCT COSTING

Product costs were calculated using a cost system developed before the introduction of the Walkman line. The primary uses of the product costs reported by this system included identifying areas for cost reduction and unprofitable features on existing products. The same basic system was used in all the firm's production facilities.

Sony's cost system distinguished between direct and indirect costs. Direct costs were those costs that could be identified with a particular product unit; indirect costs were those costs that could not be so identified.

Direct costs included purchased parts and touch labor and were measured as accurately as possible. For example, material costs were monitored by carefully listing all the purchased parts in a new product, keeping track of the selling prices negotiated with suppliers and subcontractors for these parts. Touch labor costs were monitored equally accurately by undertaking very detailed analyses of each step in the production process. For example, the time required to add the four rubber feet to a new product was carefully estimated using prior experience with other models. Where possible, these estimates were based on the actual process technology that would be used for the new product (e.g., if the rubber feet were to be screwed on instead of glued on, then the time taken to screw on feet would be used).

Indirect costs were split into two categories, those that could be associated with a specific product and those that could not. Included among the items that could be associated with a particular product were the cost of the molding and other major equipment used to produce the product and the cost of all engineering activities that could be specifically assigned to (i.e., were associated with) the product.

The assignable indirect costs were assigned to the product units by dividing the total assigned to the product by the number of units of the product that Sony expected to manufacture. The expected sales volume of a new product was estimated from the sales of its predecessor product, the degree of improvement of the new product over its predecessor, overall market trends, and feedback from retailers. Given the short product life cycles, the sales volume estimates produced by the Sony product planners were usually quite accurate.

The indirect costs that could not be assigned to individual products included R&D, engineering, service, administration, marketing, and any corporate costs charged by Sony to the general audio group. These indirect unassignable costs were allocated to each product unit as a percentage of its material cost. The same percentage was used for all products produced by the group.

CHAPTER 17

TOPCON CORPORATION

INTRODUCTION

Topcon Corporation was founded in 1932 as the Tokyo Optical Company, Ltd. The Hattori Watch Company, which subsequently became Seiko Watch Corporation, owned most of its shares. In 1947, Hattori sold its shares of Topcon in a public offering. The newly independent firm sold surveying instruments, ophthalmic refractors, and binoculars.

Topcon diversified along technological lines by relying heavily on its core competencies in advanced optics and precision equipment processing. By 1992, Topcon sold four major product lines, each treated as a separate business unit: the surveying instruments business unit contributed just under 36% of total sales; medical and ophthalmic instruments, another 28%; information instruments, 13%; and industrial instruments, approximately 23%.

This case was prepared by Professor Robin Cooper of the Peter F. Drucker School of Management at Claremont Graduate University.

Copyright © 1994 by the President and Fellows of Harvard College. Harvard Business School case 195-082.

Topcon sold only high-technology, high-margin products manufactured in relatively low volumes. Topcon management had adopted this strategy because it did not perceive that the firm had any distinctive competence to succeed in high-volume, low-margin consumer markets. The decision to specialize in low-volume, high-margin products stemmed in part from the firm's experience with manufacturing and marketing cameras, a high-volume, low-margin consumer product. The firm started manufacturing cameras in 1937 under the Lord brand name and introduced the Minion brand name in 1949. Despite being the first firm to introduce through-the-lens metering (which became the industry standard), it became clear by 1978 that the firm could not be profitable in such an intensely competitive consumer market. By 1992, virtually all camera manufacturing at Topcon had ceased.

Given its strategy, Topcon consistently had to release state-of-the-art products that matched its competitors' quality to generate the high margins required to remain profitable. This reliance on high technology and the need to develop products with increasing functionality required Topcon to invest heavily in research and development. In 1992, for example, it invested 10% of sales in research and development and also opened a ¥5 billion research and development facility.

Reflecting its strategy of heavy investment in research and development, Topcon had developed a strong reputation as a technological leader in its markets. Over the years it had introduced numerous products that were considered pioneering by both its customers and competitors. Such products included the first refractometer in 1951, the auto-level in 1961, through-the-lens metering in 1963, and the micro-theodolite in 1965. Part of Topcon's strategy was to develop a patent barrier to create a technological advantage over its competitors. While reasonably successful at this strategy, Topcon management acknowledged that Canon, one of its major competitors, was more adept at creating patent barriers.

The firm's early investment in the development of optomechatronics, a fusion of three separate technologies (advanced optics, electronics, and precision equipment processing), paid off in 1978 with the introduction of six highly successful products that relied heavily on that technology. These products immediately captured significant market share. One such product, the DM-C1 electric distance meter, was the first to use electronic distance measurement. Another was the first auto-refractometer, and a third was the retinal camera. These products differed from their predecessors by their heavy reliance on electronics that used the near-infrared portion of the electromagnetic spectrum.

The success of these products was an important milestone for Topcon. Prior to that time, the firm was in a weak financial position. For a number of years, its sales had grown slowly and profits were low. In 1976, sales were ¥7.4 billion and profits were ¥11 million. Over the next two years, sales increased by ¥1billion per year, and 1978 profits were ¥100 million. The firm proceeded to take advantage of its technological lead over its competitors by introducing a steady stream of new products, each more advanced than its predecessor. For example, in 1980 the firm revolutionized surveying instruments by combining the distance meter with the angle meter to produce the first total station, and in 1987 Topcon introduced the noncontact tonometer.

Sales and profits grew rapidly in the years after 1978. For example, from 1987 to 1991, the firm's operating income grew from just over ¥2 billion to more than ¥3.5 billion. In contrast, in the five years prior to 1987, the firm's operating income fluctuated around the ¥2 billion level.

Ophthalmic Instruments

Topcon's ophthalmic instrument line contained a number of products, each designed to perform a different set of tests. The product line included auto-refractometers, auto-kerato-refractometers, retinal cameras, and computerized lens meters. Refractometers automatically determined a patient's refractive condition (conditions tested included myopia, hyperopia, and astigmatism). Auto-kerato-refractometers provided all the usual keratometry measurements including the curvature of the cornea and its cylindrical power. The retinal cameras were used to take pictures of a patient's retina. And lens meters automatically measured the characteristics of a patient's eyeglass lenses (including the spherical power, cylindrical power, and axis prismatic power of the lenses).

The ophthalmic industry was characterized by intense worldwide competition. Topcon, the market leader with approximately 22% of the world market, recognized four significant competitors and a host of smaller ones. The largest competitor was Nidek, another Japanese firm with a 15% share of the world market, followed by Zeiss, a German firm with 10% of the world market; Humphrey, an American firm with 8%; and Roden Stock, another German firm with 6%. The other 36% of the world market was made up of approximately 50 smaller firms, including Nikon, which had a domestic market share of about 10% but a much smaller share of the world market.

Four firms dominated the Japanese auto-refractometer market: Topcon, Nidek, Canon, and Nikon. Topcon, with 29% of the domestic market, and Nidek, with 27%, were the two largest and most influential competitors. The other two firms were somewhat less influential: Canon had a

15% market share and Nikon had 13%. A number of smaller firms made up the remainder of the domestic market.

Nidek, Canon, and Nikon had entered the auto-refractometer market in the early 1980s, attracted by the high profits Topcon was earning. The three firms were able to enter the market successfully because they had already developed optomechatronic products. For example, Canon had developed photocopiers and retinal cameras, Nidek had developed laser coagulators, and Nikon had developed steppers. The entry of these three firms dramatically increased the level of competition in the ophthalmic market, causing prices to fall and profit margins to shrink. By 1992, unless a new product contained significant technological innovations, adequate profits were extremely difficult to generate. Although all the firms jockeyed for technical leadership, they were relatively evenly matched. In the early 1990s, Nidek probably had a slight advantage; however, Topcon did not consider this advantage significant.

The two market leaders, Topcon and Nidek, competed primarily on quality and functionality. The selling price of new products was set by taking into account both customers' demands and competitors' offerings. If customers were demanding increased functionality, then prices would either remain fairly constant over time or would increase. If customers wanted lower prices and were not overly concerned with functionality, then prices would fall over time. Competitive product offerings were taken into account by comparing the functionality of the various products on the market. If a new Topcon product was superior to competing offerings, then the price typically was set a little higher than competitors' prices. If the functionality was lower, then the price accordingly was set lower.

The ophthalmic business unit planning group set the selling price of Topcon's products. This group was responsible for determining the general strategic direction of each product. It would study rival product offerings, customer requirements, and technological trends before deciding on the functionality and price trends for each product. A critical task of the business unit planning group was to manage the product mix so that the overall profit equaled the target profit margin of the business unit. Typically, over the life of an auto-refractometer the price would fall slightly, usually somewhere between 5%-10%. This price reduction reflected the increased competition the product faced from newer products introduced by competitors. Because the most technologically advanced product was usually the last one introduced, the price reductions were used to offset the increased functionality of competitive offerings.

The intensified competition also led to shortened product life cycles. By 1992, new generations of auto-refractometers were introduced about every two years. Topcon management believed that two years was about the effective minimum life cycle that could be sustained for such advanced ophthalmic products. The development time for new products, combined with the amount of customer education required to introduce them, made it difficult to market new products on a more frequent basis.

In the late 1980s and early 1990s, sales of ophthalmic products by the medical and ophthalmic business unit steadily increased. This increase occurred despite the heavy competition Topcon faced. Two factors drove the growth in sales. The first was innovative products, which increased demand and enlarged the firm's market share. The second was a worldwide growth in demand for advanced ophthalmic products. This growth was driven primarily by an aging world population and increased economic prosperity in some of the most heavily populated areas of the world, such as Asia. The aging of the population increased sales because nearly everyone's eyesight deteriorated with age. The increased prosperity of certain areas of the world increased the number of people who could afford modern eye testing.

The growth in sales of advanced ophthalmic instruments was also helped by changes in the way the industry was structured. During the 1980s, there was a worldwide shift toward chain stores and away from traditional independent practitioners. Although individual practitioners typically competed on personalized service and relied on customer loyalty, the chains typically competed on price, quality of equipment, and range of services provided.

The nature of the chains and the services they provided varied by country. For example, in Japan the chains sold watches and jewelry as well as eyeglasses. These chain stores positioned eyeglasses as part of a customer's overall fashion statement. In contrast, the newest U.S. chains were located in shopping malls and offered fast turnaround, often under an hour, from prescription to glasses. Unlike their Japanese counterparts, the American chains typically sold only eyeglasses. The customers most likely to buy new models as soon as they were introduced were the chains that competed by using state-of-the-art equipment. Because these chains were growing in importance, equipment-based competition had increased the benefits of introducing new products every few years.

TOPCON'S PRODUCTION CONTROL SYSTEM

Topcon's production control system comprised two major subsystems: the turn-out-value system and the total productivity program. The latter program was in turn made up of four subsystems: zero defects, management activity by small teams, total quality control, and value analysis.

Turn-Out-Value System

Toshiba, Topcon's major shareholder, developed the turn-out-value system, or TOV. Based on an earlier General Electric design and introduced in Topcon in 1963, TOV was defined as the value designated to any kind of output from a manufacturing department including finished goods, partially finished kits, parts, or service. The TOV of a product contained three elements: direct material, direct labor, and overhead. The overall cost structure of the firm in 1992 was material, 67%; direct labor, 7%; overhead, 9%; and research and development and corporate expenses, 16%. The high material content reflected the significant level of purchased parts that the firm used in its products.

The standard direct material content of each product or component was calculated by multiplying the estimated material price (EMP) by the standard quantity of material it contained. The EMP was typically not updated during each six-month budget cycle. Only in extreme cases were standards modified to reflect price changes during the budget period. The engineering department estimated the standard quantity and included an allowance for both scrap and any cost reduction expected in the normal course of operations over the coming year. The cost-reduction or cost-down (CD) goals, as they were known, included the cost-reduction plans of the engineering, purchasing, and manufacturing departments. These plans included changes in material specifications, product design, and procurement sources.

The standard direct labor cost of each product was determined by applying the standard labor rate to the standard direct labor content of the product. The standard labor rate was the wage rate identified in the annual labor contract; this rate did not change during the year. The standard direct labor content was based on industrial engineering data from the engineering department. The standard direct labor hours were established as an attainable goal, assuming that the manufacturing, materials, and engineering departments achieved the cost-down targets for labor content reduction. The standard direct labor rate was the expected standard labor rate for the period including any anticipated wage revisions; this rate included payroll tax and other wage-related fringe benefit costs.

The standard manufacturing overhead cost for each product was determined by applying the standard manufacturing overhead rate per hour to the standard direct labor hours for the product. The standard overhead rate per hour was calculated by dividing the total standard manufacturing overhead cost for the period by the total direct labor hours for the period. The total manufacturing overhead cost was set as an attainable goal, assuming efficient expense control at standard, that is, expected, production volume for the year. The overhead expenses included the cost of indirect items such as utilities, tax, telephone, equipment and facilities depreciation, and monetary awards for the zero defects program.

The indirect costs of products were not traced to individual products in the TOV system; rather, they were calculated at the division level. For example, the medical and ophthalmic instruments division was treated as a single cost center that had only one indirect overhead rate. Despite the size of the cost center, Topcon management was satisfied that, given the relatively low level of overhead, the TOVs for ophthalmic products were sufficiently accurate for their purposes.

TOVs were calculated annually and were revised during the year only if changing conditions rendered them obsolete. The firm identified five conditions that were definite grounds for revising TOVs during the year: significant changes in material prices, in production volumes, or in the production systems or equipment; cost-reduction efforts that made standards no longer suitable goals for cost management; or changes in specification, design, or engineering that caused significant changes in the cost of products. This list was not considered exhaustive. Other, less frequently encountered conditions could require a product's TOV to be revised during the year.

The TOV system was designed to serve five primary purposes:

- profit planning (the TOV system was used to establish the firm's six-month budgets and its three-year midterm plan),
- setting the selling price of new products,
- identifying the appropriate level of price reduction in times of intense competition,
- variance analysis,
- evaluating the cost of inventory and cost of production for financial accounting purposes.

Profit Planning

The firm produced about 300 auto-refractometers per month. Weekly production was scheduled based on a monthly sales plan prepared two months previously. Topcon did not use just-in-time production but was moving toward a pull rather than push system. The production lines were tightly balanced and production was planned very precisely. Despite not using just-in-time, the firm maintained only five days' inventory across the entire line.

TOVs were used to set the six-month budgets by estimating the mix of products that were expected to be sold during that period and then multiplying the expected volumes for each product by its TOV. Revenue for the period was determined by multiplying the expected volume for each product by its anticipated selling price. Subtracting the TOV cost from the expected revenue to give the gross margin and then subtracting budgeted administration expenses and sales expenses gave the expected operating profit for the period. If the estimated profit for the period was unsatisfactory, then a cost-down plan was created to reduce the total overhead cost, labor hours consumed, and research and development expenses for the plant.

The same general procedure was used to produce the firm's three-year or midterm plans, which were used to change the firm's basic strategy. Two sets of midterm plans were developed: the static plan was used to measure how well the firm achieved its overall strategy during the three years that it was operative; the rolling plans were used to help control the firm's profits. These rolling plans were updated every year when the TOV figures were prepared. The major elements of these two sets of three-year plans were sales, cost of production (determined using the TOV system), gross margin, administration expenses, selling expenses, and operating profit.

Setting the Selling Price of New Products

The target profit margins for products were determined by competitive conditions. The process began with the average profit margin for the entire business line, set by company strategy. For example, the medical and ophthalmic business unit was treated as a single business line and expected to generate a certain percentage profit each year. This percentage was derived primarily from past experience and was set to create a difficult but not impossible target for the unit. Three factors determined the target profit margin for a given product: the relative strength of the competitive offerings, the strength of the Topcon offering, and the historical profit margins for that type of product. These three factors, combined with general business conditions, helped the ophthalmic business unit planning group set the prices of auto-refractometers.

The allowable range of market prices was determined by competitive forces. For example, Topcon would set the prices of its new products close to those of competitors' products. However, if management believed that the Topcon product had greater functionality than competitive products, then the price of the Topcon product would be higher. If the functionality was perceived to be lower, then the price would be correspondingly lower. Once the new product entered the market, competitors would react either by repricing their own products, increasing their advertising levels, or introducing a new model at a lower price. If the Topcon product was appropriately priced, it would capture the expected share of the market. If it was overpriced, market share would be lost. When this happened, Topcon management would lower its price to regain market share. No price adjustment was made when a new product's price was found to be too low. Following standard business practice in Japan, prices were not raised even if a product was extremely successful and raising its price was likely to increase the overall profits of the firm.

Identifying the Appropriate Level of Cost Reduction

The TOV system was an integral part of the firm's drive to reduce costs. Although TOVs anticipated expected cost-reduction efforts, the system created pressure for the workers to achieve even greater savings by calculating the income from production. This income number was calculated by subtracting the actual cost of production from the TOV cost for the same product mix. If the income from production was positive, then the efficiency of the firm was higher than expected; if negative, then it was lower. When the income was lower, the business unit was expected to find ways to improve performance and bring the income number back under control.

In addition to creating these two pressures to reduce costs, the TOV system was used to identify products that were rapidly losing profitability and required more focused cost-reduction efforts. Products lost profitability as their selling prices fell to compensate for the higher functionality of new products introduced by competitors. Although some cost reduction was possible across the life of the product, the short life cycles coupled with high material content made it extremely difficult for Topcon to reduce costs as rapidly as prices fell. One of the advantages of short product life cycles was that they held price reductions to a minimum.

Topcon had two types of cost-reduction programs: planned and spontaneous. Both these programs were considered benefits of the TOV system. Planned projects were initiated before production began; spontaneous programs were initiated after it began. Examples of planned programs included increasing parts commonality and converting die-cast parts to plastic. Spontaneous programs consisted of the firm's continuous improvement programs, the daily suggestion schemes, and the zero defect system. Examples of spontaneous programs included renegotiating prices with vendors, changing the source of a component from external to internal supply, and reducing run times.

Doing Variance Analysis

Variance analysis was used to create pressure on the work force to reduce costs. Monthly variance analysis reports, prepared by the accounting division, were sent to all the firm's production sites. The major variances calculated were direct material usage and price variances, direct labor rate and efficiency variances, and indirect expense spending and volume variances. These variances were accumulated over the six months of the current budget.

The variances were calculated using the usual formulas. The only unusual feature was that costs associated with research and development and headquarters charges were deducted from indirect expenses before the variances were calculated.

Evaluating the Cost of Inventory

The TOVs were used to determine the value of inventory. Topcon's tightly linked production system required that only minimal inventory levels and TOV costs be used to determine the cost of inventory and hence the cost of goods sold for financial accounting purposes.

Total Productivity Program

Topcon management did not believe that variance analysis alone led to cost reduction. Although the TOV system included a number of different variances, they were all based on the difference between TOV and actual value. According to management, the results of such analyses were "just a set of numbers." To reduce costs, the activities performed and the products produced needed to be linked. Topcon called this link "total productivity." Thus, though the TOV system was intended to create pressure to reduce costs, the total productivity program was designed to reduce costs in the firm's production facilities.

In 1988 Topcon got a new president, Masayoshi Motoki, who had come from Toshiba. He initiated the New Topcon Program, which was intended to improve corporate culture and strengthen corporate capability. The major shift in the corporate culture envisioned by the new president was to make the firm more aggressive in the marketplace. In particular, Motoki wanted the firm to be more aggressive in the way it priced its products and in the way it designed and marketed them. The New Topcon Program consisted of general capability improvements that focused on sales and marketing, technological capability improvements that dealt with the acquisition of advanced manufacturing technologies, and total productivity efforts that sought to improve product delivery times, quality, and cost. In 1992, the program was expanded and renamed the Best Topcon Program (TP).

The TP program formalized and integrated the four existing cost-reduction programs—zero defects (ZD), management activity by small team (MAST), total quality control (TQC), and value analysis (VA). The TP program itself applied only to production areas because it had an underlying objective, the introduction of lean manufacturing, including just-in-time production. In contrast, the four constituent programs were applied throughout the firm.

Zero Defects Program

The zero defects program consisted of groups of 10 to 15 workers who tried to find ways to improve either the way they performed their daily work or the part of the production process they managed.

MAST Program

The MAST program consisted of cross-functional, self-directed teams formed to achieve a specific objective. A typical objective for an accounting MAST team would be to find ways to improve the accuracy of the inventory records. The equivalent objective for a MAST production team would be improving inventory control.

MAST teams operated at the division manager level. The leader of the team would form a cross-functional team, usually from manufacturing, production, accounting, marketing, and, in particular, the TQC and VA departments. Typically, each manager would be a leader of one team and a member of several others. This interlocking membership allowed the teams to be aware of the actions of other teams and thereby avoid duplication of effort.

Total Quality Control

The total quality control program was designed to improve the quality of the daily work performed in each section of the firm, including sales, production, design, and administration. The program integrated the zero defects program with MAST and factory quality initiatives.

Value Analysis

Value analysis (VA) was used to reduce product costs at the design stage. A VA committee in each business unit, such as ophthalmic and medical instruments, surveying instruments, and industrial instruments, was mandated to find ways to design costs out of products, for example, by replacing expensive components with less expensive ones and by eliminating unnecessarily demanding specifications. The committees prepared a monthly report that detailed what actions they had taken and the results of those actions. These monthly reports were then accumulated into a company-wide report.

VA techniques used a hierarchical management structure. The business unit or plant manager was considered the promoter of VA. The area manager was responsible for the VA project. The section manager had the task of implementing the VA proposals, and the VA task force had to achieve the savings. The many different types of VA projects; their frequency, and some examples are shown in Table 17-1. Most of the projects were design related.

Under VA, when the actual cost was greater than the TOV, three actions were typically taken to reduce cost.

• Reduce material cost by redesigning the product to reduce the material content (either by changing the type of material involved or by changing the quality of material consumed).
• Negotiate a better price for the material.
• Decrease the run time.

VA programs typically were initiated during the development of a new product. The process began with defining the general concept of the product and its place in the existing product line. The ophthalmic business unit planning group would determine the strategy that the product was following (e.g., increased functionality and increased price versus constant functionality and decreasing price) and would decide the appropriate price and functionality for the product. After the product design was approved, a prototype was constructed and the product's expected TOV or target cost was determined using cost tables. If the computed ex-

TABLE 17-1. TYPE AND FREQUENCY OF VALUE-ADDED PROJECTS

Type of Project	Frequency	Examples
Product design	62.1%	Convert die-cast to plastic injection Increase parts commonality Reduce the size of a display CRT Cost down
Purchased parts	8.6%	Negotiate price reduction with vendors
Efficiency improvement	5.5%	Reduce run times
Structural modification	4.7%	Eliminate components by product redesign
Equipment modification	3.9%	Customize equipment Reduce fixturing requirements
Other	2.9%	
Process improvements	2.5%	Improve efficiency of scheduling
Software	2.1%	Increase commonality of software
Miscellaneous	7.7%	Jig improvement Kaizen proposals Change in purchased parts Parts elimination
Total	100.0%	

pected profit margin of the product was unsatisfactory, a value analysis project was initiated.

Great care was taken to initiate only beneficial VA projects. Management was particularly concerned that the pressure created by the firm's direct-labor-based cost system to reduce a product's direct labor content might lead to projects that were not cost beneficial. In particular, care was taken not to initiate projects that reduced the direct labor content of a product, but not overhead, if the projects cost more than the direct labor they saved. The initiation of such projects was avoided by justifying all VA projects at the project, not the product, level.

The cost structure of Topcon's products, with their high material and nonmanufacturing content, made it difficult for the production facilities to have much effect on total product costs. The facility-level controllable costs were relatively small compared to total costs, and the plants were already efficiently managed. Consequently, Topcon management believed that the most effective way to reduce costs was to design new, less expensive products.

CHAPTER 18

ISUZU MOTORS, LTD.

INTRODUCTION

Isuzu Motors, Ltd. (Isuzu) traces its origins back to 1916, when the Tokyo Ishikawajima Shipbuilding and Engineering Co., Ltd., began to manufacture automobiles. The division merged with Dat-Auto in 1933 and changed its name to Motors, Inc. In 1937, it merged with Auto Manufacturing, a division of Tokyo Gas and Electricity, Inc., to form Tokyo Motors. The firm's name was changed to Isuzu Motors in 1949.

In 1992, Isuzu was the ninth-largest automobile company in Japan, based on units produced. Isuzu recognized 10 domestic competitors. The largest four were Toyota, with a 32% market share; Nissan, with 17%; Mitsubishi, with 11%; and Mazda, with 10%. The other six players shared

---

This case was prepared by Professor Robin Cooper of the Peter F. Drucker School of Management at Claremont Graduate University and Professor Takeo Yoshikawa of Yokohama National University.

the remaining 30%. Honda, Suzuki, Daihatsu, and Fuji Heavy Industries were all larger than Isuzu, which had a 4% market share. Only Hino and Nissan Diesel were smaller, with a 0.7% and 0.4% market share, respectively. Isuzu's market share was somewhat misleading because it averaged out the firm's strong market position in trucks and buses with its weak position in the higher-volume passenger car market. The same was true of both Hino and Nissan Diesel.

Isuzu had 1992 sales of ¥1,200 billion, up from ¥769 billion in 1984, making it the sixth-largest Japanese automobile firm in terms of sales. During that period, both the percentage of vehicles exported and the mix of vehicles sold remained fairly constant. Exports accounted for approximately 49% of sales, while heavy-duty trucks accounted for 23%, light-duty trucks and buses for 34%, and passenger cars for 14%; parts, engines, and components accounted for the remaining 29% of sales.

Isuzu's market share in heavy- and light-duty trucks was just over 10%, making it the fourth largest in the domestic Japanese market. Its major competitors were Toyota, Mitsubishi, and Nissan, with market shares of 24%, 14%, and 11%, respectively. Isuzu had a slightly higher share of the bus market—11%—which made it the fifth-largest Japanese bus manufacturer. Its major competitors were Toyota, Mitsubishi, Hino, and Nissan, with market shares of 26%, 24%, 19%, and 13%, respectively. Only in light trucks was Isuzu the market leader, with a market share of 35%.

With respect to medium- and heavy-duty trucks (exceeding 6.1 tons in gross vehicle weight), Isuzu had a 33% market share in Japan in 1992, and was the second-largest manufacturer in the world after Mercedes-Benz. Isuzu's share in Japan was also 33% for 2-ton trucks, which was by far the largest sector of the Japanese truck market. In fact, until 1993, Isuzu had been the top-ranking Japanese firm in this sector for 23 consecutive years.

Production was done at four domestic plants. The oldest plant was located in Kawasaki, 10 miles from Tokyo. The facility manufactured heavy- and light-duty trucks and buses as well as heavy-duty industrial engines. The Kawasaki facility was built in 1937 and for 25 years was Isuzu's sole production facility. After 1962, the firm began to expand aggressively and opened other manufacturing facilities. The Fujisawa plant, also located in the Kanagawa prefecture, was built in 1962. It produced light-duty trucks, buses, and passenger cars as well as light-duty industrial engines and components.

The Tochigi plant was located in the Kanagawa prefecture, some 50 miles from Tokyo. It was built in 1972 to produce engines and drive-line components for the firm's product line. Later, the plant was expanded to cope with Isuzu's increased sales volume. The Hokkaido plant was located in the Hokkaido prefecture, about 800 miles north of Tokyo. It was built in 1984 to manufacture engines for passenger cars and had the capacity to build approximately 200,000 engines a year. Many of these engines were sold to other automobile manufacturers, both domestic and international, including Opel (General Motors' European subsidiary) and General Motors in the United States.

In 1991, Isuzu was unprofitable for the first time in many years. Although several factors contributed to the poor results, the main problem was that Isuzu's passenger vehicles failed to maintain market share. Reacting to these poor sales, the firm adopted a four-part strategy. First, it increased its efforts to sell recreational vehicles, which were popular in Japan and were highly profitable; increased sales of these vehicles would help overall profitability immediately. Then, to maintain its position in the bus market, Isuzu increased its rate of new product introductions. Third, it revamped its line of heavy-duty cars; Isuzu was the world leader in heavy-duty cars, followed by Hino and then by Mercedes-Benz. Isuzu's objective for the heavy-car market was to increase its range of heavy-duty cars and to update its product line. Finally, top management was reviewing its North American strategy. Isuzu and Fuji Heavy Industries had formed a joint venture to sell their products in North America. The joint venture had performed poorly in recent years, and Isuzu was studying options for a new approach to selling automobiles in North America.

COST-CREATION PROGRAM

Isuzu's small size (compared to most of its competitors) required the firm to maintain a constant downward pressure on costs. It was difficult to achieve significant cost reductions after a product was designed and ready for mass production (particularly after the rapid appreciation of the yen). Changes that late in the design process required reevaluating the performance, functionality, and quality of the product. Such changes were extremely expensive to make and were therefore often uneconomical; to achieve significant cost reductions, the product had to be designed with cost in mind. Therefore, creating a mechanism to introduce cost-reduction techniques as early as possible in the planning process was considered critical if the cost-reduction program was to be successful.

All cost-reduction programs, or cost-creation programs, as they were known, were the responsibility of Yoshihiko Sato. Sato had joined Isuzu in 1964 and was responsible for much of the progress the firm had made in successfully applying target costing and value engineering approaches. He

EXHIBIT 18-1. RESUME OF YOSHIHIKO SATO

Work Experience: Isuzu Corporation, 1964–present

Garuda Corporation (on temporary transfer from Isuzu)

Cost Planning Division: Spread cost management and VE in Isuzu and subcontractors.

Product Development and Management Section: Spread cost-kaizen technique in Isuzu and subcontractors.

Isuzu Motors, Ltd.–Manager

Development and implementation of the teardown method in Isuzu domestic and overseas factories (1972–present).

Study and implementation of VE in Isuzu and other firms. Gave guidance in production techniques for 15 years in Japan and other foreign countries.

Licenses

Certified Value Specialist by the American Society of Value Engineering

Memberships

Counselor, Japanese Society of Value Engineering

Vice President, Tokyo Chapter of Japanese Value Engineering Society

Instructor, Kansai Chapter of Japanese Value Engineering Society

Special Member, Isuzu Kyowa-kai, Cost Committee

Special Member, Isuzu Kyowa-kai, VE Committee

Publications

Books and papers on cost management.

was highly respected in value engineering circles and had held many senior positions in several of the Japanese value engineering associations (for a brief resume, see Exhibit 18-1).

Sato preferred the term "cost creation" over "cost reduction" because he felt that it suggested a more active perspective. To Sato, cost reduction suggested that the product was reduced in the process. Sato did not want people to think that the way to reduce costs successfully was simply to reduce a product's functionality. Instead, he wanted them to realize that the real task was to provide exactly the same functionality but at a lower cost. At the heart of the firm's cost-creation program were two critical documents: the cost deployment flowchart and the cost strategy map.

Cost Deployment Flowchart

To help achieve its cost-reduction objectives, Isuzu developed a sophisticated approach to cost creation that relied heavily on a cost deployment flowchart (see Figure 18-1). The cost deployment flowchart was developed to ensure that cost-reduction activities were applied to a product as early as possible in its development. The flowchart identified five development stages in a vehicle's design: the concept proposal stage, the planning

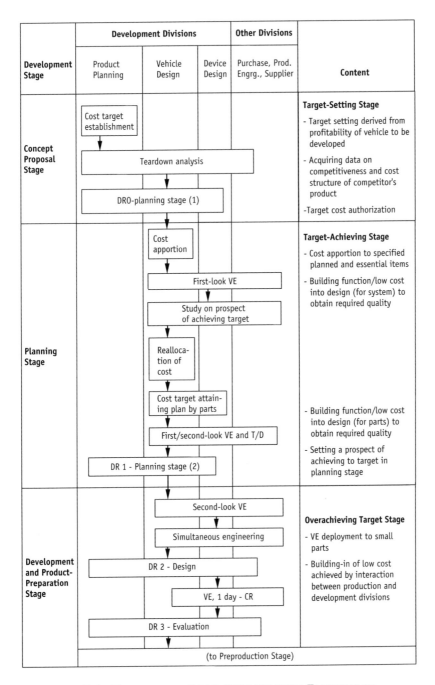

	Development Divisions			Other Divisions	
Development Stage	Product Planning	Vehicle Design	Device Design	Purchase, Prod. Engrg., Supplier	Content
Concept Proposal Stage	Cost target establishment → Teardown analysis → DRO-planning stage (1)				**Target-Setting Stage** - Target setting derived from profitability of vehicle to be developed - Acquiring data on competitiveness and cost structure of competitor's product -Target cost authorization
Planning Stage		Cost apportion → First-look VE → Study on prospect of achieving target → Realloca-tion of cost → Cost target attaining plan by parts → First/second-look VE and T/D → DR 1 - Planning stage (2)			**Target-Achieving Stage** - Cost apportion to specified planned and essential items - Building function/low cost into design (for system) to obtain required quality - Building function/low cost into design (for parts) to obtain required quality - Setting a prospect of achieving to target in planning stage
Development and Product-Preparation Stage		Second-look VE → Simultaneous engineering → DR 2 - Design → VE, 1 day - CR → DR 3 - Evaluation			**Overachieving Target Stage** - VE deployment to small parts - Building-in of low cost achieved by interaction between production and development divisions
	(to Preproduction Stage)				

FIGURE 18-1. PORTION OF COST DEPLOYMENT FLOWCHART

stage, the development and product-preparation stage, the development and production/sales-preparation stage, and the production/sales-preparation stage.

The flowchart was developed after Isuzu management realized that no significant analysis had been performed during the product-planning stage to ensure that a product's target cost could be achieved. When it finally became clear that the product could not be manufactured for its target cost, it was too late in the development process to effectively and economically reduce its cost sufficiently to achieve the target cost.

The cost deployment flowchart also identified three major elements of the cost-reduction program:

- The different types of cost-reduction activities and the development stages at which they would occur. For example, during the concept proposal stage a new vehicle's target cost was determined, teardown analysis was performed, and a design review was held.

- The development division responsible for the cost-reduction activity. For example, while the product program planning division was solely responsible for establishing the target cost, almost every division was responsible for the teardown analysis.

- The contents of the cost-reduction activity. For example, it described the steps required to achieve a product's target cost.

Cost Strategy Map

Isuzu had developed many cost-reduction techniques, including several types of value engineering and teardown methods. To ensure consideration of appropriate cost-reduction techniques at the right time in the development process, a cost strategy map was developed (see Figure 18-2). This map was required because the appropriate technique depended both on the stage of the development process and the part in question. For example, some cost-reduction techniques, such as first- and second-look value engineering, were suitable only for making drastic changes to the design of a part. They were not suitable when an existing design was being slightly modified.

The cost strategy map followed the cost deployment flowchart; for each development stage, the applicable cost-reduction techniques were identified. For instance, for the planning stage, the cost deployment flowchart identified two major blocks of cost reduction: first-look value engineering and a mixture of first- and second-look value engineering and

○ Responsible Section
● Related Section

Stage	Development process	Completion date	Things to be done (target value)	Tool to be used	Manual No.	Department responsible for implementation and its role: Product Planning Office – Planning	Product Planning Office – Cost Control	Product Planning Office – Cost Planning	Design – Vehicle	Design – Device	Experiment	Production Engineering	Purchase	Suppliers	Content
Planning Stage	Cost target achievement plan for individual parts		Preparing overall execution plan and commonization plan							○					
			Preparing achievement plan for the individual tools	Part strategy list	Format exists			●		○			●		Cost tradeoff possible within device
	Setting target for individual parts		Investigation of parts to be procured abroad							○			●		Investigation of actual record of rival models
			Preparing VE plan							○					Submitting VE register to REG
			Preparing T/D plan	T/D theory	VEE-185	○			○	●					Carried out mainly with parts for which VE is not done
			Preparing 1-day CR plan	1-day CR	VEA-103								○		Carried out with C/O parts (basic design is not changed)
			Judging the possibility			●		●	●	○			●		Determining cost for main parts individually and in groups
	Activities for achieving cost target		Carrying out VE	1st & 2nd-look VE	VEE-158	●		●		○	●	●	●		
			Specifying clearly the functions of the VT candidates	VT	VEE-166	●		●	●	○					
			Carrying out T/D	D-T/D, C-T/D	VEA-172	●		●		○	●	●	●		Carrying out dynamic T/D, cost T/D
			Guessing the cost of each part					○		●		●	●		
			Guessing the investment for each part					○		●		●	●		
			Preparing matrix T/D chart	Matrix T/D	VEE-170					○		●	●		M-T/D chart for parts for commonization to be updated regularly
			Wiring, piping, and other layout						○	●	●	●			

FIGURE 18-2. COST STRATEGY MAP

							Product Planning Office	Design				Production Engineering			
						◯ Responsible Section										Department responsible for implementation and its role
						● Related Section										
Stage	Development process	Completion date	Things to be done (target value)	Tool to be used	Manual No.	Planning	Cost Control	Cost Planning	Vehicle	Device	Experiment	Production Engineering	Purchase	Suppliers		Content
Development and Production-Preparation Stage	Activities for achieving cost target		Carrying out VE	1st-look VE 2nd-look VE Mini-VE	VEE-158 VEE-056 VEE-159	●		●		◯	●	●	●			When rough picture of big assemblies emerges, VE is deployed for smaller part picture. Mini-VE is done for parts that are even smaller
			Carrying out ordinary CR activities	VE checklist	VEA-121		●		◯			●	●			A list of unreflected items is prepared after checking
	Simultaneous engineering		CR activities for overall profitability							◯						Extending CR activities to major parts of vehicles other than the control vehicle (selection of part and selection of person responsible for its promotion), verification of product weight (preparation of a comprehensive weight list, procuring estimate)
			Incorporating production department's views on assembly of prototype vehicle							◯		●				Check to see whether production department's views on prototype vehicle have countermeasures prepared if any problems are found

FIGURE 18-2. COST STRATEGY MAP *(continued)*

teardown. The corresponding cost strategy map identified seven techniques: three types of teardown methods and four types of first- and second-look value engineering.

Together the cost deployment flowchart and the cost strategy map were used to ensure that the appropriate cost-reduction techniques would be applied as soon as possible in the development of new products. To maximize cost reduction, Isuzu created a comprehensive cost-reduction program that consisted of many different cost-reduction techniques. Isuzu developed some of these techniques and adopted others from systems developed at other firms.

COST-DOWN PROGRAM

Isuzu's cost-down program consisted of five major steps: developing functional specifications, value engineering, teardown, detailed cost reduction, and some cost-reduction-related techniques, such as the Isuzu Production System (IPS) and design for manufacturability.

The IPS was a modified form of just-in-time production that allowed suppliers to hold specified amounts of inventory. It was considered different from JIT because it included the improvement of all production activities, not just inventory, for example, analyses and improvements in the distance workers had to move during their jobs. The distinction was necessary because in Japan the term JIT had been expanded to include such activities, and everyone meant something different by it. To avoid confusion, Isuzu used the term IPS to describe its use of JIT and associated production-improvement techniques.

Developing Functional Specifications

There were two cost-reduction programs in the functional specification stage, the symbiosis research system (SRS) and the Mona Lisa program. Professor Hiroshi Tsuchiya of the Sanno Institute of Management in Tokyo developed the SRS program. It was first implemented at Matsushita, the giant Japanese electronics firm. The objective of SRS was to find rational (i.e., more efficient) manufacturing techniques through the analysis of daily work habits. At the heart of SRS was the systematic questioning of everyday activities. Research had shown that most people took their daily tasks so much for granted that they never bothered to analyze them to see if they could be either performed more efficiently or eliminated. This "blindness" created a natural barrier to applying kaizen (continuous improvement) effectively to such activities. SRS was designed to overcome this blindness by creatively analyzing work habits and eliminating unnecessary movements.

The objective of the Mona Lisa program was to make a product the best of its type ever made by integrating the overall product development program. The program was named after the painting by Leonardo Da Vinci, considered by many one of the finest paintings of all time.

Value Engineering

Isuzu's value engineering program had eight aspects: value target, zero-look value engineering, first-look value engineering, second-look value engineering, manufacturing engineering,the wave method, mini-value engineering, and the value engineering reliability program.

Value Target

Value target (VT) was Isuzu's term for the procedures it had developed to identify the target costs of components purchased from outside suppliers. As part of the planning stage, the target cost for an entire vehicle in the concept proposal stage was distributed among the vehicle's 8,000-10,000 components at the major function or group component levels. Isuzu designers identified approximately 30 major functions per vehicle, including the engine, transmission, cooling system, air conditioning system, and audio system. Group components were the major subassemblies purchased from the firm's suppliers and subcontractors. There were only about 100 such components, yet they amounted to as much as 70%-80% of the manufacturing cost. Group components included the carburetor and starter.

The target costs of these major functions and group components were determined using monetary values or ratios. The monetary values were determined by using market research that asked customers to estimate how much they would pay for a given function. Ratios were developed by asking customers to estimate the relative importance of each function using a 100-point scale. The monetary values and ratios were used to prorate a product's target cost to the major functions. Although customer analysis was the primary mechanism for prorating target costs, factors such as technical, safety, and legal considerations often led to adjustments to the prorated target costs. For example, if the prorated target cost for a component was too low to allow a safe version to be produced, the component's target cost was increased and the target cost of the other components was decreased.

Once the target costs were prorated, Isuzu contacted its outside suppliers to begin the bidding process. The nature of the bidding process depended on the strength of a supplier's relationship with the firm. For some suppliers, Isuzu would provide capital and design support to enable them to produce the advanced components required for the next generation of products. For other firms, the relationship was less important and the supplier offering the greatest value was chosen; for these types of bids, three suppliers were contacted for each component and asked to develop prototypes. The suppliers were told the target quality, functionality, and price of the component and were expected to produce prototypes that satisfied all three requirements.

Once the three prototypes were submitted, Isuzu engineers would analyze them to determine which one provided the best value. Value was defined as the functionality of the component divided by its cost. Functionality was more than simply whether the part could perform the task for which it was designed; it also included the convenience and prestige of the design. This expanded definition of functionality was adopted when consumers began to demand variety and luxury in their products, not just functionality. No numerical analysis was performed when determining a part's value, only a subjective one. The product with the highest perceived value was selected and the component's supplier was typically awarded the total contract.

Although the supplier rated as having the highest value would generally win the order, firms that had a reputation for being good suppliers were often awarded at least part of the order even if their products did not have the highest value. Examples of such companies included Yuasa for batteries, Toyo Valve for valves, and Nihon Seiko for bearings. These firms were awarded partial contracts to maintain their relations with Isuzu.

The primary benefit of this "maximize value" approach was its ability to bring out the strengths of each supplier. When ethical, Isuzu would share innovations made at one supplier with other suppliers to help them achieve target costs and make an adequate return. Originality also played a role in the selection of the winning design. If one of the suppliers found a way to add additional functionality to the component that increased its value, Isuzu engineers would incorporate this functionality into the part's specifications. Typically, the creative supplier would achieve a higher value than the other suppliers because its component already contained the extra functionality. In this way, suppliers were encouraged to act as ancillary research and development laboratories for Isuzu. Incorporating the additional functionality spread the innovation to other suppliers, thus increasing their abilities to provide higher-value-added components in the future.

Zero-Look Value Engineering

Zero-look value engineering was the application of value engineering techniques to the earliest stages of product development. It was performed during the concept proposal stage, during which a product's preliminary quality, cost, and investment targets were established. Traditionally, value engineering started with the first look; around 1980, Isuzu developed its zero-look program, which made the concept of designing revolutionary products an integral part of product design. Through zero-look value engineering, Isuzu's engineers were expected to find revolutionary solutions to improve the functionality of the firm's products. Identifying the potential of automatic transmissions when only manual ones were available is an example of the type of innovation fostered by zero-look programs.

Zero-look value engineering was applied in the development of the firm's NAVI-5 transmission system. This revolutionary system combined the benefits of a manual transmission with those of an automatic one. In simplified terms, NAVI-5 was a computer-controlled manual transmission capable of changing gears automatically. The advantage of the new transmission was its ability to combine the convenience of an automatic transmission with the higher fuel efficiency and performance of a manual transmission. Zero-look value engineering identified the basic concept behind this new system.

The objective of zero-look value engineering was to find ways to add value to products from the consumer's perspective while simultaneously increasing the firm's portion of the value-added to the product. The NAVI-5 system achieved both objectives, based on market studies that showed consumers liked the performance provided by a manual transmission but found the process of changing gears in Japan's crowded cities too tiring. Consumer preference testing indicated a strong interest in the hybrid transmission. The new product increased the portion of Isuzu's value-added because Isuzu manufactured its manual transmissions (and purchased automatic ones from a supplier). Without the discipline of zero-look value engineering, the firm probably would not have identified the NAVI-5 as a new product.

First-Look Value Engineering

First-look value engineering—defined as developing new products from concepts—was applied during the last half of the concept-proposal stage and during the entire planning stage. In the planning stage, the key components or major functions were identified; the commodity value (i.e., the product's type, quality, size, price, and function) was determined; a design plan was submitted; the target costs were distributed to the vehicle's major functions, such as its engine, transmission, and air conditioner; and the degree of component commonality was set. First-look value engineering was the application of value engineering principles as the blueprints were drawn. The objective was to increase a product's value by increasing its functionality without a corresponding increase in cost.

Isuzu's engineers applied first-look value engineering to the development of the Gemini heater. A conventional automobile heater functioned only when the vehicle's engine was warm. In winter the occupants would remain cold until the engine temperature increased sufficiently. Users took this inconvenience for granted. First-look value engineering identified that reducing the time it took for the automobile interior to warm up would be a welcome benefit for users. Consequently, a project was initiated to find ways to heat the car interior before the engine warmed up. The ultimate solution was to install a ceramic heater that functioned only when the engine was below a specified temperature. This heater was used to warm the air flow directed to the occupants' feet. When the water in the engine reached the specified temperature, the ceramic heater switched off and the traditional heater took over.

Second-Look Value Engineering

Second-look value engineering was applied during the last half of the planning stage and the first half of the development and product-preparation stage. In the development and product-preparation stage, the components of the main functions were identified and the first hand-made prototype was assembled. Unlike zero- and first-look value engineering, the objective of second-look value engineering was to improve the value and functionality of existing components, not to create new ones. Once improved, these components were incorporated into Isuzu's new products.

An example of the application of second-look value engineering was the folding gear lever on the ELF, a light-duty truck. Experience with earlier models had shown that the gear lever, which was positioned between the two front seats, sometimes got in the way of the occupants. The vehicle's functionality and hence its value would be improved if the gear lever was repositioned so that it was out of the way. The solution was to develop a gear lever that could fold down while the vehicle was stationary but would not collapse while the vehicle was in motion.

Manufacturing Value Engineering

The objective of manufacturing value engineering was to identify the best approach to produce a part. The critical trade-off was quality versus cost. The functionality of the product had already been determined. Manufacturing value engineering was applied during the last half of the development and product-preparation stage and the first half of the development and production/sales-preparation stage. During the development and production/sales-preparation stage, the detailed parts drawings were updated and the second prototype was produced. Typical applications of second-look value engineering included selecting the way a component was cast and deciding whether joints should be manually or robotically welded. Such applications took into account both technology and economics when deciding which approach to use.

Wave Method

The wave method was a mechanism to incorporate value engineering techniques systematically into small-group activities such as quality control and industrial engineering. It was applied on a continuous basis during the development and product-preparation stage, the development and production/sales-preparation stage, and the production/sales-preparation stage. The wave method used a "working group approach": each group would analyze problems encountered with new products and use functional analysis to identify ways to improve the design.

Mini Value Engineering

Mini value engineering was a simplified approach to second-look value engineering. It was applied to specific areas of a part or to very small, inexpensive parts, such as mirrors, doors, and door locks. A typical outcome of mini value engineering would be mirrors that were more ergonomically designed. Like the wave method, mini value engineering was applied during the development and product-preparation stage, the development and production/sales-preparation stage, and the production/sales-preparation stage.

Value Engineering Reliability Program

The value engineering reliability program was designed to ensure that the most appropriate form of value engineering was applied to each problem. For example, if a completely new product design was required, then applying second-look value engineering would not be appropriate; either first- or zero-look value engineering would be required. Like the wave method and mini value engineering, the value engineering reliability program was applied during the development and product-preparation stage, the development and production/sales-preparation stage, and the production/sales-preparation stage. This program was especially useful for the value engineering of critical safety-related parts.

Teardown

Teardown was a method used to compare Isuzu's products and components with those of its competitors. Fundamentally, the purpose of teardown was to analyze competitors' products in terms of materials, parts, function, manufacture, coating, and assembly. Isuzu defined the teardown method as "a comparative value analysis method through visual observation of disassembled equipment, [with] parts and data arranged in a manner convenient for such observation." Teardown was used in all stages of product development.

Teardown methods were introduced in 1972, the year after General Motors (GM) purchased a 37% share of Isuzu. Sato was exposed to GM's teardown approach for about 30 minutes and realized that GM's approach was the basic teardown approach documented in the literature. The GM method, or static teardown method, as it was known in Isuzu, was subsequently modified over a three-year period to fit Isuzu's needs and then allowed to proliferate throughout the company. The Isuzu teardown method attracted considerable attention and became the basis for what is now known as the Japanese teardown method.

The primary difference between the original GM approach and Isuzu's method was the scope of the application of teardown principles. Isuzu's teardown program contained eight different teardown methods: dynamic, cost, material, static, process, and matrix teardown, plus the unit-kilogram price method and the group estimate by teardown method. The first three methods were designed to reduce the direct manufacturing cost of a vehicle. The next three sought to reduce the investment required to produce vehicles via increased productivity. The last two methods were integrations of teardown and value engineering techniques.

Dynamic Teardown

The objective of dynamic teardown was to identify ways to reduce either the number of assembly operations required to manufacture a vehicle or the time it took to perform them. The method consisted of taking apart competitors' products and analyzing their assembly processes to compare them with Isuzu's and see if their application to Isuzu's products would be beneficial.

Cost Teardown

The objective of the cost teardown method was to reduce the cost of the components used in a vehicle. A cost comparison of the components used by Isuzu and its competitors was performed. If an Isuzu component was more expensive, its functionality was analyzed to see if the extra cost provided increased functionality. If the functionality was the same, the component was subjected to cost-reduction activities, such as functional improvement, weight reduction, reduced setup times, and reduced number of processes.

Material Teardown

The material teardown approach consisted of comparing the materials and surface treatments of the components that were used by Isuzu and its competitors. Competitive products would be purchased and torn down so that any innovations introduced by competitors could be identified and, if thought advantageous, adopted by Isuzu for future products. To be effective, the analysis was restricted to parts with the same work function. For example, in a given subassembly, Isuzu might be using metal parts while its competitors might be using less expensive plastic ones.

Static Teardown

The static teardown approach was the most basic of all the teardown approaches. It consisted of disassembling a competitor's product into its components and then laying them out on a table or putting them up on a display board so that the design engineers could see the difference between their products and those of their competitors.

Process Teardown

Process teardown consisted of comparing the manufacturing processes for similar parts and reducing the difference between them, with the long-term objective of producing multiple products or components on the same production line. Achieving this objective was particularly important to Isuzu because its relatively small production volumes required mixed production if it was to be profitable. Reflecting this objective, Isuzu was the only automobile manufacturer that could produce mixed products, such as four-door, two-door, and left- and right-hand-drive cars, on the same assembly line.

Matrix Teardown

In the matrix teardown method, a matrix was developed of all components used in Isuzu products. This matrix, prepared on an as-needed basis, identified the volume of each component used per month by each model and the total usage across all models. Any low-volume components were flagged, actively designed out of existing products, and banned from future products.

Unit-Kilogram Price Method

In the unit-kilogram price method, parts that were produced using similar manufacturing processes were treated as a product group and analyzed for possible cost savings opportunities. In this approach, the efficiency of the product or component was expressed in terms of its value per kilogram (determined by dividing its cost by its weight).

Products that required further analysis were identified by plotting the value per kilogram for all the products in the same product group against their weight. The resulting scatter diagram was used to identify the outliers that had much higher values per kilogram than the majority of products. These outliers were analyzed to determine why they had such a high cost per kilogram, and methods were identified to reduce their material cost. The targeted savings were identified as the difference between a product's current cost and its cost at the group's average value.

Group Estimate by Teardown Method

The group estimate by teardown (GET) method was a combination of basic value engineering and teardown procedures; it was a modified version of the unit-kilogram price method. The GET method consisted of treating as a group parts that had similar functions and analyzing them for possible cost savings. For example, a windshield washer tank and a radiator surge tank both performed the same fundamental function: holding liquids. Because the two tanks performed essentially the same function, under the GET method they were compared to see if there were ways to make them more efficiently. For example, in some designs the two tanks were combined into a single tank with two compartments.

A scatter diagram of the cost of the products in each group versus their functionality was constructed to help identify products with abnormally high costs. Typically, abnormalities were caused by differences in the product's design, the production methods used, production volumes (as identified by the teardown matrix), and the level of external purchases. In this comparison, the reasons behind these differences were analyzed and appropriate cost-reduction actions taken. For example, if the cause of the high costs was the use of special materials, the part was redesigned to allow for more conventional materials to be used. Similarly, if the production volume was low, ways were found to standardize the manufacture of the parts so high-volume production could be used.

Detailed Cost Reduction

In addition to the techniques used to develop functional specifications, value engineering, and teardown, Isuzu relied on detailed cost-reduction techniques that were used when, after the application of value engineering, a part still could not be produced at its target cost. These techniques included the checklist method and the one-day cost-reduction meeting.

Checklist Method

The checklist method used a checklist of questions to identify possible methods to further reduce the cost of a product or part. Issues such as number of parts, shape, surface treatment, and the accuracy demanded were all analyzed to see if they could be modified to save costs. A typical checklist is shown in Exhibit 18-2. One of the advantages of the checklist method was that it helped ensure that all possible avenues for cost reduction were explored.

One-Day Cost-Reduction Meetings

The objective of the one-day cost-reduction meetings was to improve the efficiency of the entire cost-reduction activity, including the application of value engineering and teardown methods. At these meetings, attended by experts from engineering, production, cost, and sales, participants were expected to identify cost-reduction possibilities that had not been explored to date. The meetings were designed to overcome limitations in the *ringi* approval process used for most cost-reduction proposals. Ringi were written proposals circulated to all involved parties; individuals signed off on a

EXHIBIT 18-2. TYPICAL CHECKLIST

1. **Number of parts**
   a. Can you reduce the number of parts?
   b. Is there any possibility that you can reduce costs by increasing the number of parts?

2. **Shape**
   a. Can you make it smaller?
   b. Can you make it lighter?
   c. Can you make it simpler and more standardized?
   d. Is there any different shape that makes it easy to process?
   e. Is it suitable for the next process to produce?
   f. Is it a suitable shape to increase yield ratio?

3. **Materials**
   a. Can you change quality of materials?
   b. Can you get more inexpensive materials?
   c. Is it easy to process?

4. **Surface treatment**
   a. Is it a suitable surface treatment?
   b. Is surface heat treatment suitable enough?

5. **Accuracy**
   a. Is it easy to achieve given accuracy?
   b. Is it more accurate than expected?

6. **Processing and assembling**
   a. Can you manage more units of machines?
   b. Can you integrate production processes?
   c. Can you save more production time?
   d. Can you reduce number of Kosuu?
   e. Can you save set-up time?
   f. Is it an optimal lot size?
   g. Can you speed up more?

7. **Equipment**
   a. Can you make it smaller?
   b. Do you have more inexpensive equipment?
   c. Can you make it automatic?

8. **Models**
   a. Can you make it smaller?
   b. Can you make it simpler?
   c. Is the quality of materials good enough?
   d. Does it take advantage of old models?

9. **Workers**
   a. Do you have an ineffective movement?
   b. Is the workforce optimal?
   c. Is the work process standardized?

10. **Inspection and shipping**
    a. Is the defect rate high?
    b. Do you have useless shipping?

proposal to accept it. The problem with this approach was that it limited face-to-face interactions where ideas could be freely exchanged and modified. The results of various teardown programs were used at the meeting to help initiate discussion.

Each one-day meeting produced three results. New cost-reduction ideas were conceived, actual cost-reduction actions were identified, and a follow-up schedule was developed to ensure that the actions identified at the meeting were actually implemented.

One-day meetings were held either when a new product's cost was still above its target cost after the last cost-reduction step or when economic conditions were particularly adverse and the firm needed to save money on the manufacture of existing products.

APPLYING THE TECHNIQUES

The integration of the cost deployment flowchart and the cost strategy map created a flexible and powerful cost-reduction program. Although each cost-reduction technique could be used in several stages of the product development process, most were designed to be used at specific stages.

Sato discussed his views on barriers to the success of Isuzu's cost-creation program:

> The first barrier is caused by the mind-set of our employees. The type of analyses that underlie cost creation are fundamentally simple. The greatest problem we face is that people do not go back to the basics and miss opportunities. Therefore, our greatest challenge is to find ways to make people identify the opportunities that exist.

> The second barrier is caused by the lack of knowledge of our employees. While there are many opportunities for cost creation, not enough of our employees understand value engineering and other cost-reduction techniques. Consequently, they do not know what is possible and therefore, while they may see an opportunity, they do not know how to take advantage of it. We try to reduce this problem by educating as many people as possible.

> The third barrier is caused by timing problems. Unlike production processes that are either running or broken down, costs tend to grow slowly. While no one ignores a production process that has broken down, people often ignore increased costs because they hope somebody else will take care of it. People only do anything when the problem has become so large that it cannot be ignored.

> With the current downturn we have increased the size of our cost creation teams significantly. The original team contained seven highly

trained members. They were called the "brain team": they came up with the ideas and others implemented them. In December 1992, we added 23 new members to the cost creation team. In October 1993, we added another 22 members to this second team. The two teams have been very active finding ways to reduce costs. In the first two months the team identified savings worth ¥2.2 billion. Their target for next year is ¥8.4 billion.

COMPANY DESCRIPTIONS

Citizen Watch Company, Ltd.

This company was the manufacturing arm of the world's largest watch producer, Citizen, founded in 1930. It was not only responsible for manufacturing watches but was also strategically diversified into products that required expertise in watch technology: numerically controlled production equipment, flexible disk drives, liquid crystal displays for television and computers, dot matrix printers, and jewelry. The non-watch products consisted of almost half its revenues in 1990.

Higashimaru Shoyu Company, Ltd.

This manufacturer of soy sauce was formed in 1942 by the merger of the Kikuichi Shoyu Goshi Gaisha and Asai Shoyu Gomeri Gaisha. Higashimaru produced light and dark soy sauce (80 types), Japanese-style porridge, Japanese-style salad dressing, sweet sake, soup stocks, and noodle sauces. This variety of products, with the light soy sauce being most important, generated approximately ¥21 billion of sales in 1992 and employed 510 people.

Isuzu Motors, Ltd.

Isuzu originated in 1916 with Tokyo Ishikawajima Shipbuilding and Engineering Co., Ltd., manufacturing automobiles. Based on units produced, Isuzu was the ninth-largest automobile company in Japan in 1992 with 10 large domestic competitors. The 4% market share of Isuzu was misleading because it did not reflect the firm's specialized market strength in trucks and buses due to the higher-volume passenger car market. Isuzu had 10% market share in heavy-and light-duty trucks and 11% share in the bus market.

JKC

JKC was founded in 1955 when it became independent of Diesel Kiki Co., Ltd. Its products were directly related to the three basic functions of a vehicle: driving, turning, and stopping. They included brakes, clutches, steering systems, and pumps. Among its major customers were Isuzu, Toyota, Nissan, and Komatsu. Its annual production was approximately ¥22 trillion.

Kamakura Iron Works Company, Ltd.

Kamakura, founded in 1910 as a blacksmith company, was a family-run firm located in a distant suburb of Tokyo. The firm has remained relatively small with 1993 sales of nearly ¥6 billion and profits of ¥35 million. The firm has been a supplier of automotive parts, with 21 major customers, including Yokohama Corporation (40% of sales), Isuzu Motors (20%), Hino Motors (15%), Jidosha Kiki Company (10%), and Yamaha Motors (5%). The majority of its customers have been either automobile manufacturers or suppliers to that industry. Other customers have included Iseki, Kayaba Industries, and Shinryo Heavy Equipment. Although large portions of the revenues were from vertically integrated companies, Kamakura was an independent company and did not belong to a keiretsu.

Kirin Brewery Company, Ltd.

Kirin originated as the Spring Valley Brewery in Yokohama in 1870, when W. Copeland, an American, established Japan's first brewery. The firm has diversified into a wide range of products—biotechnology-based pharmaceutical products, new hybrid vegetable varieties, and optical sensing systems—using its beer manufacturing technologies. They also produced products derived from by-products of the brewing process such as carbonated drinks, yeast-related feed for fish and livestock, and yeast-derived natural seasonings. With these products, Kirin generated ¥1,800 billion and ¥90 billion in sales and operating profits respectively.

Komatsu, Ltd.

Founded in 1917 as part of the Takeuchi Mining Co., this firm was one of the largest heavy industrial manufacturers in Japan. It was organized into three major lines of business—construction equipment, industrial machinery, and electronic-applied products—which accounted for 80% of total revenues. The remaining 20% consisted of construction, unit housing, chemicals and plastics, and software development. These products together generated revenues of ¥989 billion and net income of ¥31 billion, making Komatsu a large international firm. Since 1989, the company has been aggressively diversifying and expanding globally.

Kyocera Corporation

Kyocera was founded in 1959 by its chairman, Dr. Kazuo Inamori, and seven of his colleagues as the Kyoto Ceramics Company, Ltd. It considered itself to be a "producer of high-technology solutions," specializing in developing innovative applications of ceramics technology. Kyocera established a strong reputation for innovation among major technical leaders in the semiconductor and electronic industries by taking on technically impossible jobs. The firm had sales of ¥453 billion and net income of ¥27 billion in 1992.

Mitsubishi Kasei Corporation

Formerly known as Mitsubishi Chemical Industries, Ltd., this company was Japan's largest integrated chemical company with ¥710 billion in revenues and ¥5 billion in net income. Its three major groups were carbon and inorganic chemicals, petrochemicals, and functional products. The first two groups consisted of high-volume and mass-produced products, whereas the functional products were relatively low in volume with high value-added. The firm had successfully implemented the strategy to diversify by adding functional products to the firm's traditional product offerings.

Miyota Company Ltd.

Miyota Co. Ltd. (Miyota), founded in 1959, was a 100%-owned subsidiary of Citizen Watch Company, Ltd. In 1995, its sales were just over ¥36 billion. The firm was originally dedicated to the assembly of watches for its parent, but over the years it diversified along technology lines by specializing in miniature mechatronic (the fusion of mechanical and electronic technologies) products. By 1995, Miyota produced four major product lines: completed watches, watch movements and parts, quartz oscillators, and viewfinders for camcorders.

Nippon Kayaku

Juntaro Yamamoto in 1916 founded this first manufacturer of industrial explosives in Japan. In 1992, the firm had five major lines of business—pharmaceuticals, sophisticated products, agrochemicals, dyestuffs, and explosives and catalysts—that were organized into pharmaceuticals and fine chemicals. Nippon Kayaku had sales of ¥117 billion and net profit of ¥2.8 billion in 1992. The firm's growth was achieved both through internal expansion and through several acquisitions and mergers.

Nissan Motor Company, Ltd.

This firm, founded in 1933, considered itself the most highly globalized of the Japanese automobile companies with 36 plants in 22 countries and marketing in 150 countries through 390 distributorships and over 10,000 dealerships. In 1990, Nissan was the world's fourth-largest automobile manufacturer, producing just over 3 million vehicles, about 10% of the world's demand for cars and trucks. Nissan had a stated policy of globalization through a five-step process that emphasized localization of production, sourcing, research and development, management functions, and decisions.

Omachi Olympus Ltd.

Omachi Olympus Co., Ltd. (Omachi) was a 100%-owned subsidiary of Olympus Optical Co., Ltd. (Olympus). It specialized in producing complex, curved, plastic moldings primarily for incorporation into the camera products that were made by the Consumer Products Division of its parent. The firm was located in the city of Omachi, which was in the Nagano Prefecture some 150 miles from Tokyo.

Olympus Optical Company, Ltd.

As part of Olympus, Olympus Optical Company manufactured and sold opto-electronic equipment and other related products. Originally Takachiho Seisakusho, Olympus was founded in 1919 as a producer of microscopes.

Major product lines were cameras, video camcorders, microscopes, endoscopes, and clinical analyzers. By 1995, Olympus was the world's fourth-largest camera manufacturer, with consolidated revenues of ¥252 billion and ¥3 billion in net income.

Shionogi & Co., Ltd.

This firm was founded as a wholesaler of traditional Japanese and Chinese medicines by Gisaburo Shiono in 1878. Shionogi was a research and development-oriented pharmaceutical manufacturer with 12.4% of sales dedicated to R&D. Shionogi's strategy focused on selling its products to hospitals and universities. The majority of its revenues were generated through pharmaceutical products; however, it also had other business such as animal health products, agro-chemicals, industrial chemicals, diagnostics, and cosmetics. It was recognized around the world for the quality of its antibiotics and other pharmaceutical products with ¥225 billion in sales in 1992.

Sony Corporation

Sony, one of the world's largest electronics companies, started as Tokyo Telecommunications Research Institute and generated revenues by repairing broken radios and manufacturing short-wave converters in its earlier years. The company's first really successful product was Japan's first magnetic tape recorder in 1950. The company continued to grow rapidly, and by 1960 it became a truly international firm with Sony Corporation of America and Sony Overseas, S.A., in Switzerland, followed by Sony UK and Sony GmbH in 1968 and 1970 respectively.

Sumitomo Electric Industries, Ltd.

This firm was founded in 1897 as a manufacturer of bare copper, Sumitomo Copper Rolling Works. Since then, Sumitomo Electric Industries (SEI) has continued to produce electric wires and cables and was the world's third-largest manufacturer of these products. The firm has adopted a diversification program since 1931, taking advantage of the distinctive competencies in the manufacturing of electric wires and cables. SEI's top management considered the firm to be one of the most highly diversified in Japan.

Taiyo Kogyo Co., Ltd. (The Taiyo Group)

This firm was founded in 1947 by Kuniyasu Sakai and Hiroshi Sekiyama as The Taiyo Painting Company. As the firm grew and the Japanese economy expanded, it diversified from painting into metal stamping and electronic equipment. The Taiyo Group's Bunsha philosophy was to have small companies with highly autonomous managers to avoid bureaucracy. This philosophy was based on Sakai's notion that "when a company gets too large it cannot respond in time." He focused on the flexibility of the small firms as a powerful form of cost management.

Topcon Corporation

Topcon was originally founded as the Tokyo Optical Company, Ltd. in 1932. It diversified along its core competencies in advanced optics and precision equipment processing. By 1992, Topcon sold four major product lines: surveying instruments, medical and ophthalmic instruments, information instruments, and industrial instruments. The surveying instruments business unit contributed approximately 36% of sales; medical and ophthalmic instruments, 28%; information instruments, 13%; and industrial instruments, 23%. Topcon specialized in high-technology, high-margin, low-volume products. To rely continuously on high technology for profit, Topcon invested heavily in research and development.

Tokyo Motors Ltd.

Tokyo Motor Works, Ltd. (TMW), when measured in terms of worldwide production, was by 1990 one of the world's top 10 automobile manufacturers. The firm, which was founded in 1945, by 1990 produced just over 2 million vehicles, supplying approximately 4% of the world's demand for cars and trucks. Of these vehicles, slightly over 1.2 million were passenger cars. TMW produced vehicles at 20 plants in 15 countries and marketed them in 110 countries through 200 distributorships and more than 6,000 dealerships.

Toyo Radiator Co. Ltd.

Toyo Radiator Co. Ltd. (Toyo) was founded in 1936 as a radiator supplier to the fledgling Japanese automobile industry. Toyo was not associated with and was independent from any of the major kieretsus. Over the years it diversified into all the arenas of heat-exchange applications, and by 1995 it sold heat-exchange products for use in automobiles and heavy construction and agricultural vehicles, air conditioners for home and office, and freezers. Its product lines included radiators, oil coolers, intercoolers, evaporators, and condensers. In 1995 it was one of the world's largest independent heat-exchange equipment manufacturers for construction equipment.

Toyota Motor Corporation

Toyota Motor Corporation (Toyota) started as a subsidiary of the Toyoda Automatic Loom Works, Ltd. It was founded in 1937 as the Toyota Motor Company, Ltd. It changed its name to the Toyota Motor Corporation in 1982 when the parent company merged with Toyota Sales Company, Ltd. In 1993 Toyota Motor Corporation was Japan's largest automobile company. It controlled approximately 45% of the domestic market. Over the years, Toyota had changed from a Japanese firm into a global one. In 1993 a considerable part of the firm's overseas markets were serviced by local subsidiaries that frequently designed and manufactured automobiles for local markets.

Yamanouchi Pharmaceutical Company, Ltd.

This firm was founded in 1923 to manufacture and sell pharmaceutical products. In 1992 Yamanouchi was Japan's second-largest pharmaceutical drug company in terms of net profit and was highly respected both in Japan and internationally. A. T. Kearney designated this company as one of the best performing companies in the world in 1990. This company had a strong corporate philosophy of "Creating and Caring . . . for Life." In 1992 the consolidated sales were ¥357 billion and the net income was ¥33 billion; pharmaceutical products accounted for 73% of sales, nutritional products, 17%, and food and roses, 10%. In 1989 as part of its diversification strategy, Yamanouchi acquired the Shaklee Group, where the bulk of its nutritional products were sold.

Yamatake-Honeywell Company, Ltd.

Founded in 1906 as a small trading company, this firm has grown into a group of six companies with consolidated 1993 sales of ¥196.8 billion and pretax income of ¥14.5 billion. In 1993, the group consisted of Yamatake & Co., Ltd., Yamatake Keiso Co., Ltd., Yamatake Engineering Co., Ltd., Yamatake Control Products Co., Ltd., Yamatake Techno-systems Co., Ltd., and Yamatake-Honeywell Co., Ltd. Yamatake-Honeywell comprised four divisions that carried out research and development for control and automation: industrial systems, building systems, control products, and factory automation systems.

Yokohama Corporation, Ltd.

Yokohama was founded in July 1939 as a joint venture between a Japanese automobile manufacturer and the Japanese government. The objective of the firm was to manufacture hydraulic systems for automobiles and trucks and associated equipment under license from a German firm. Firm ownership changed over the years and in 1993 only three major shareholders remained: Isuzu Motors, Nissan Motors, and The Industrial Bank of Japan. By 1992, the firm had 13 overseas affiliates, seven liaison offices, and a worldwide service network of over 100 distributors and 2,000 service representatives with sales of ¥257 billion and 6,800 employees. Yokohama was split into three corporate divisions: injection pump, air conditioning, and hydraulics and pneumatics.

GLOSSARY OF TERMS

ACTIVITY-BASED COSTING (ABC). A highly accurate product-costing methodology that assigns costs to products based on the activities required to produce the product. ABC traces costs to products using many different activity measures that reflect the quantities of input consumed to manufacture the product.

ALLOWABLE COST. The difference between the target selling price and the target profit margin.

AMOEBA MANAGEMENT SYSTEM. A system developed at Kyocera that organizes the firm into a large collection of quasi-autonomous profit centers or highly independent pseudo firms responsible for selling a number of products both internally and externally.

AS-IF COST. The cost of a future product if it were manufactured today, taking advantage of cost-reduction opportunities identified when the previous generation of products was being designed or manufactured.

BATCH-AND-QUEUE. An approach to task completion that consists of making a large number of similar items and then queuing that batch before the next step in the process.

BENCHMARK ALLOWABLE COST. The allowable cost of the product if it was designed and manufactured by the most efficient producer in the industry.

BUNSHA PHILOSOPHY. An approach to managing firm size used by the Taiyo Kogyo Group founder Kuniyasu Sakai, who believes strongly that small firms are inherently more efficient and effective than large firms. Bunsha, meaning "to divide," refers to the spinning off of new venture firms from the parent firm.

CARDINAL RULE OF TARGET COSTING. Rule that says that the target cost can never be exceeded.

CONFRONTATION. A competitive strategy that assumes sustainable, product-related competitive advantages are unlikely to be developed. Confronting competitors head on is achieved by rapidly matching the cost/quality/functionality improvements other firms initiate while nurturing the ability to create temporary competitive advantages.

COST-DOWN PROGRAMS. A disciplined companywide cost-reduction campaign with a primary focus on finding ways to design costs out of products before they enter production. These programs also include methods for improving efficiency in the manufacturing process.

COST LEADERSHIP. A generic competitive strategy that requires the firm to establish itself as the lowest cost producer in an industry. Traditionally, cost leaders offer products that are low in price and functionality.

COST-REDUCTION OBJECTIVE. The difference between the allowable cost and the current cost.

CURRENT COST. The cost of a future product, assuming that it is manufactured today from existing components using existing production processes.

DESIGN FOR MANUFACTURE AND ASSEMBLY (DFMA). A simultaneous engineering process that optimizes the relationship between materials, manufacturing technology, assembly process, functionality, and economics. It seeks to ease the manufacture and assembly of parts or eliminate them.

DIFFERENTIATION. A generic competitive strategy that relies on the provision of unique product offerings that closely satisfy customers' requirements. The differentiator's product offerings are usually high in both functionality and price.

DRAFT TARGET COST. The expected cost of manufacture arrived at through target costing methods. The draft target cost is an updated expected cost arrived at after computing the projected cost of each major subcomponent. It is used primarily to determine whether or not a product can be produced and sold profitably.

DRIFTING TARGET COST. The estimate of what the product cost would be if the product were manufactured at any point during the product design process given confirmed cost savings.

EXPECTED COST. The cost determined by product designers who have no specified cost objective to achieve but are expected to minimize the cost of the new product as it is designed; used in cost-plus pricing approaches.

EXPECTED SELLING PRICE. The price used in the product pricing methods associated with a cost-plus approach. It becomes the dependent variable and is determined by adding a target profit margin to the expected product cost.

FINAL TARGET COST. The target cost set at the very end of the product design stage. Unlike earlier draft target costs, it includes both direct and indirect manufacturing costs.

FIRST-LOOK VALUE ENGINEERING. A technique that focuses on the major elements of the product design. Its objective is to enhance functionality by improving the capability of existing functions.

FUNCTIONAL GROUP MANAGEMENT SYSTEM. The Olympus Corporation's term for a bottom-up approach to identifying and achieving cost-reduction targets that are carried out by self-directed work teams.

INTERORGANIZATIONAL COST MANAGEMENT SYSTEM. A cost-reduction program initiated by a core firm and carried out across the entire value chain. These systems contribute to the blurring of organizational boundaries as firms share information and resources to improve the efficiency of interfirm activities.

JUST-IN-TIME (JIT). The ordering and delivery of parts as they are needed in the production process to achieve minimum inventory and waste.

KAIZEN. A Japanese term that stands for continuous improvement. As applied in cost management practice, the term refers to a total commitment on behalf of the work force to finding new ways to reduce costs and increase efficiency in the manufacturing process.

KAIZEN COSTING. The application of kaizen techniques to reduce the costs of existing components and products by a prespecified amount. Its objective is to reduce a product's cost through increased efficiency of the production process.

KEIRETSU. A federation of firms joined in a relationship based on their common traditions and business dealings. Keiretsu can be vertical, between a manufacturer, its component/material suppliers, and its distribution partners, or horizontal, usually centered around a financial institution that holds equity stakes in the participants while it meets their financing needs.

LEAN ENTERPRISE. A new organizational form originating in Japan. It employs lean production methods such as just-in-time production, total quality management, team-based work arrangements, supportive supplier relations, and improved customer satisfaction. The lean enterprise is capable of producing high-quality products economically in lower volumes and bringing them to market faster than mass producers.

MASS PRODUCER. A firm that relies on mass production manufacturing techniques similar to those pioneered by Henry Ford, among others.

MAXIMUM ALLOWABLE PRICE. The highest price the customer is willing to spend for the product irrespective of its quality and functionality.

MAXIMUM FEASIBLE VALUES (of a survival triplet characteristic). Determined by the capability of the firm, maximum feasible values are the highest values for product characteristics a firm can achieve without violating the minimum or maximum values of the other characteristics.

MINIMUM ALLOWABLE VALUE (of a survival triplet characteristic). Determined by the customer, the minimum allowable level is the lowest value of each product-related characteristic that the customer is willing to accept.

OPERATIONAL CONTROL. Method for obtaining feedback on how well costs are being controlled. Assigning responsibility to individuals for costs they control makes responsibility centers the primary unit of analysis to evaluate how well costs are controlled, through the use of variance analysis.

PERFECT WASTE-FREE COST. The cost at which a product could be manufactured if no waste were created in the production process.

PRODUCT COSTING. A systematic cost accounting method for assigning appropriate costs to products during the manufacturing process to determine product costs and monitor profitability.

QUALITY FUNCTION DEPLOYMENT. A visual decision-making procedure for multiskilled project teams. It develops a common understanding of the voice of the customer and a consensus on the final engineering specifications of the product that has the commitment of the entire design team.

RESERVE FOR PRODUCTION MANAGER. The difference between the product-level target cost and the sum of component-level target costs, assembly, and indirect manufacturing target costs. It is used to allow for minor cost overruns in the production process that are caused by design problems.

SECOND-LOOK VALUE ENGINEERING. Applied during the last half of the planning stage and the first half of the development and product preparation stage, its objective is to improve the value and functionality of existing components.

SINGLE-PIECE FLOW. An approach to task completion that consists of completing each product or design in its entirety without any queues.

STRATEGIC COST-REDUCTION CHALLENGE. The difference between the allowable and product-level target costs.

SURVIVAL TRIPLET. Three product-related characteristics—cost (price), quality, and functionality—that must be managed to ensure that products remain within their survival zones.

SURVIVAL ZONE. Area defined by the gaps between minimum and maximum price, functionality, and quality. Only products that fall inside their survival zone can be successful.

SUSTAINABLE COMPETITIVE ADVANTAGE. The ability to sustain product-related or other competitive advantages and thus generate high profit margins for a lengthy period of time.

TARGET COSTING. A structured approach to determining the cost at which a proposed product with specified functionality and quality must be produced to generate the desired level of profitability at its anticipated selling price. A product's target cost is determined by subtracting its target profit margin from its target selling price and adjusting for the strategic cost-reduction challenge.

TARGET COST-REDUCTION OBJECTIVE. The difference between the current cost and the target cost.

TARGET PROFIT MARGIN. The target profit margin is set based on corporate profit expectations, historical results, competitive analysis, and in some cases computer simulations. Target profit margins applied in target-costing procedures ensure that products will be sold at a minimum acceptable profit.

TARGET SELLING PRICE. The selling price of a new product determined primarily from market analysis. It is used to determine a target cost, which is applied during the design phase of new product development. It is also used as the basis for determining the purchase price of components and raw materials acquired externally.

TEARDOWN ANALYSIS. A method used to analyze competitive product offerings in terms of materials and parts used as well as ways in which they are manufactured.

TEMPORARY COMPETITIVE ADVANTAGE. An advantage achieved over one's competitors that is not expected to last for long.

TOTAL QUALITY MANAGEMENT (TQM). An integrated approach that focuses on designing quality into products and ensuring that the production process is as defect free as possible.

UNAVOIDABLE WASTE-FREE COST. The cost at which a product could be manufactured if no unnecessary waste was created during the production process.

VALUE ENGINEERING (VE). A systematic interdisciplinary examination of factors affecting the cost of a product so as to devise a means of achieving the required standard of quality and reliability at the target cost.

ZERO DEFECTS (ZD). A quality program with the objective of reducing defects to zero.

ZERO-LOOK VE. The application of VE principles at the concept proposal stage, the earliest stage in the design process. Its objective is to introduce new forms of functionality that have not previously existed.

SELECTED BIBLIOGRAPHY
AND REFERENCES

Selected Bibliography on Target Costing

Brausch, J. M. "Beyond ABC: Target Costing for Profit Enhancement." *Management Accounting*, November 1994, pp. 45–49.

This practitioner article describes how an American textile manufacturer implemented a target costing program to reduce the costs of its products at the design stage and improve the firm's profitability. The evolution of cost management at the textile firm is described. First, the firm separated the management accounting function from financial accounting. Second, the cost system was redesigned using ABC principles to obtain more accurate product costs. The third step focused on reducing costs proactively through target costing. The goal was to enhance the firm's profits through concentrated cost-reduction efforts and continuous improvement in product design. The article emphasizes the importance of using a cross-functional, team-based approach to target costing and the need to integrate cost management with the other activities of the firm.

Cooper, R. *When Lean Enterprises Collide: Competing Through Confrontation*. Boston: Harvard Business School Press, 1995, 379 pp.

This book introduces a new theory of competition, the confrontation strategy, in which sustainable competitive advantages cease to exist and lean enterprises become locked in relentless, head-on competition. These firms engage in a competitive game of constantly introducing new products with the desired levels of quality and functionality at the lowest cost. Cooper shows that the key to success in such a competitive environment is the careful balance of the three elements in the "survival triplet"—price/cost, quality, and functionality—with an emphasis on aggressive cost management across the supply chain. Eight integrated cost management techniques are described, including target costing and value engineering. The crucial role that these techniques play in supporting the confrontation strategy is demonstrated using vignettes from practice. The evidence on which the insights in this book are based is drawn from an in-depth field study of the strategy and cost management practices inside 20 Japanese firms.

Cooper, R., and W. B. Chew. "Control Tomorrow's Costs through Today's Designs." *Harvard Business Review*, January-February 1996, pp. 88–97.

Cooper and Chew emphasize the need for feedforward cost management in the new global economy, where lean enterprises with their fast reaction times, quickly erode first-mover advantages and enter into intense competition on price, quality, and functionality. Target costing is described as a highly disciplined management process designed to ensure that the competitive pressures from the marketplace get transmitted to the product designers and to create a common language for the various participants in the development process. The article shows how two Japanese firms, Olympus Optical Company and Komatsu, have successfully developed and used target costing systems to survive in the harsh environment of lean competition and to put pressure on their suppliers to reduce the costs of externally acquired parts.

Fisher, J. "Implementing Target Costing." *Journal of Cost Management*, Summer 1995, pp. 50–59.

This article describes the target costing process at two Japanese companies, Matshushita Electric Works and Toyota Motors. Both companies use target costing for cost reduction at the product-planning and product-design stages. The article discusses the various stages in the target costing process, including value engineering, and shows the similarities and differences between the two firms. A number of difficulties in implementing target costing are also highlighted.

Hiromoto, T. "Another Hidden Edge—Japanese Management Accounting." *Harvard Business Review*, July-August 1988, pp. 22–26.

In this article Hiromoto discusses how Japanese companies use their management accounting and control systems to reinforce their strategies. In particular, he emphasizes how Japanese manufacturers incorporate a strong commitment to market-driven cost management and continuous improvement into their management accounting systems. To illustrate this point, Hiromoto describes the target costing process at Daihatsu Motors. Another observation is that standard costing is not used as widely in Japan as it is in Western companies. Instead, target costs are set based on market demands and planned profit levels and are tightened over time.

Horvath, P. *Target Costing: A State-of-the-Art Review.* IFS International Ltd., Bedford, U.K., 1993, 64 pp.

This monograph provides a summary of a literature review highlighting some important topic areas relating to target costing. It defines the objectives and scope of target costing and briefly describes the process of setting target costs. The relationship between target costing and management accounting, including activity-based costing and life cycle costing, is discussed. The monograph describes how target costing is embedded in the product design and planning processes of the firm and the role of the firm's marketing function and its suppliers. Special tools for target costing addressed in the monograph include target cost matrices, value control charts, and cost tables. The review concludes with a summary of survey results on Japanese practice.

Kato, Y. "Target Costing Support Systems: Lessons from Leading Japanese Companies." *Management Accounting Research*, Vol. 4, No. 4, 1993, pp. 33–47.

In this article, Kato explores the contribution of target costing to cost management in Japanese firms. He provides a number of reasons why attention has shifted from just-in-time production and cost reduction at the manufacturing stage toward target costing, which focuses on the product planning and development stage. Information systems necessary to support the target costing process are described, including support systems for market research, target profit computation, R&D, and value engineering (e.g., cost tables). The article also warns against potential disadvantages of target costing, such as the extreme demands it places on the work force.

Kato, Y., G. Boer, and C. W. Chow. "Target Costing: An Integrative Management Process." *Journal of Cost Management*, Spring 1995, pp. 39–51.

This article addresses the organizational aspects of target costing implementation and some of the dysfunctional effects of target costing. Two case studies of target costing processes at Japanese companies, Daihatsu Motors and Matshushita Electric Works, are presented. Central themes of the article are the profit management orientation of target costing and target costing as an integrative management system that links the various functional areas of a business. The authors discuss potential downsides of target costing: longer development cycles, employee burnout, market confusion, and organizational conflicts.

Monden, Y. *Cost Reduction Systems. Target Costing and Kaizen Costing.* Portland, OR: Productivity Press, 1995, 373 pp.

This book contains a comprehensive and highly structured description of the target costing and kaizen costing processes. Monden provides considerable detail on the different stages in the process, the various methods and systems used, and the role of the departments responsible for each step in the process. Examples from the automotive industry are included to illustrate practice. Part I describes the 14 basic steps of target costing with a chapter devoted to each step. Part II presents more detail, with chapters on how target costing works in division-based companies, how companies undertake target costing semiconcurrently with their parts manufacturers, how companies can implement target costing companywide, and how they can use value engineering. Part III provides details of the cost estimation system, with specific information on the target costing approach to estimating direct materials costs and processing cost rates, estimating labor-hours using time tables, and combining the main cost elements in cost estimation for new products. Part IV describes kaizen costing.

Monden, Y., and K. Hamada. "Target Costing and Kaizen Costing in Japanese Automobile Companies." *Journal of Management Accounting Research*, Vol. 3, Fall 1991, pp. 16–34.

In this article, Monden and Hamada describe the features of total cost management systems observed in Japanese automobile companies, including target costing and kaizen costing. The term "total" implies cost management in all phases of a product's life (i.e., target costing systems to support the cost-reduction process in the product development and design stage and kaizen costing systems for the manufacturing stage) as well as total involvement of all people in all departments. The features of target costing and the various steps in the target costing process are described. The second part of the article contrasts kaizen costing with traditional standard costing and explains the computation and assignment of kaizen cost-reduction objectives.

Monden, Y., and M. Sakurai, eds. *Japanese Management Accounting: A World Class Approach to Profit Management.* Cambridge, MA: Productivity Press, 1989, 546 pp.

This book contains a collection of articles written by different authors on various topics related to Japanese management accounting. The material dealing with target costing and/or value engineering includes: "Recent Trends in Japan's Cost Management Practices" by T. Makido; "Cost Planning and Control Systems in the Design Phase of a New Product" by M. Tanaka; "A Japanese Survey of Factory Automation and Its Impact on Management Control Systems" by M. Sakurai and P. Y. Huang; and "Total Cost Management System in Japanese Automobile Corporations" by Y. Monden.

Morgan, M. J. "Accounting for Strategy: A Case Study in Target Costing." *Management Accounting (CIMA)*, May 1993, pp. 20–24.

This article criticizes the inappropriateness of traditional standard costing for cost-control purposes in the modern business environment and advocates target costing as a market-driven approach that links operational control to strategy. Morgan gives a practical example of how the target cost is computed for a customer order and which actions can be taken to achieve the cost target, such as reducing the number of defects and tooling costs, and using simultaneous engineering.

Sakurai, M. "Target Costing and How to Use It." *Journal of Cost Management*, Summer 1989, pp. 39–50.

This article is one of the earliest writings in English explaining target costing and its uses. Sakurai describes the economic and social factors that led Japanese companies to adopt target costing. He defines target costing, identifies its properties, and describes the basic steps in the target costing process. The article also discusses a number of management tools used in the practical implementation of target costing, such as just-in-time production, value engineering, and total quality control. Target costing is said to be most effective in assembly-oriented industries, which are characterized by high product proliferation and increasing capital intensity. Sakurai suggests cost management for software development as a new area where target costing can be successfully applied.

Society of Management Accountants of Canada, Institute of Management Accountants, and Consortium for Advanced Manufacturing-International. *Implementing Target Costing*. Management Accounting Guideline No. 28, April 1994, 49 pp.

This monograph defines the concept of target costing and describes the steps in the process of implementing target costing. It presents a set of tools and techniques to facilitate the target costing process, including tools for market assessment, competitive analysis, and financial planning, cost tables, value engineering, and quality function deployment. The monograph also discusses the role of management accountants in the target costing process and how they can be key contributors to the introduction and successful application of target costing.

Tanaka, T. "Target Costing at Toyota." *Journal of Cost Management*, Spring 1993, pp. 4–11.

This article illustrates how target costing is used at Toyota to reduce costs at the product-design stage. Toyota pioneered target costing in Japan, and its current approach to target costing is widely considered one of the most advanced. Using cost planning, Toyota sets goals for cost reduction (based on the difference between target profit and estimated profit) and then seeks to achieve those goals through design changes. The article describes the firm's unique differential approach to cost planning, based on summing the estimated differences in cost between the new and current models. Once the chief engineer has distributed the cost-reduction goals to the design divisions, value engineering is undertaken to help achieve those goals. The article describes the process used to monitor the evolution of costs as a result of design changes and at the mass-production stage.

Tani, T., H. Okano, N. Shimizu, Y. Iwabuchi, J. Fududa, and S. Cooray. "Target Cost Management in Japanese Companies: Current State of the Art." *Management Accounting Research*, Vol. 5, 1994, pp. 67–81.

This article explores the range of industries that have adopted target cost management, the organization and tools of target costing, the cooperation with suppliers, and the transfer of target costing to foreign subsidiaries. The research is based on a mail survey conducted in 180 Japanese companies. The results indicate that target costing is applied predominantly in assembly and process industries and that it is implemented to serve multiple objectives including cost reduction, quality assurance, timely introduction of new products, and product development to satisfy customer needs. Product development by cross-functional teams and simultaneous engineering are identified as critical elements of the target costing process. Common practices identified in setting target costs are adjustment of the allowable costs to reflect the forecasted actual costs[1] and decomposition of target costs by part. In addition, the achievement of target costs is found to be com-

---

[1] Forecasted actual costs have been identified as "as-if costs" in this volume.

monly monitored throughout the process and even during mass production. Cost tables and value engineering case collections are identified as typical tools for target cost management. Collaboration with suppliers is also identified as one of the critical factors for the successful transfer of target costing to foreign subsidiaries.

Yoshikawa, T., J. Innes, F. Mitchell, and M. Tanaka. *Contemporary Cost Management*. London: Chapman & Hall, 1993, 187 pp.

This book discusses several Japanese and Western approaches to cost management. The techniques covered include target costing, activity-based cost management, and strategic management accounting. The authors describe the various steps in the target costing process and provide practical guidelines for companies considering the implementation of the technique. The primary focus of the book is on a product life cycle perspective and the information that management accounting can provide in a proactive manner as opposed to the more traditional reactive one. Chapters 4 (Target Costing), 5 (Functional Analysis), 6 (Cost Estimation), and 7 (Cost Tables) are of particular interest for individuals thinking of adopting target costing.

Other Readings on Target Costing

Baker, W. M. "The Missing Element in Cost Management: Competitive Target Costing." *Industrial Management*, March/April 1995, pp. 29–32.

Bhimani, A., and H. Okano. "Targeting Excellence: Target Cost Management at Toyota in the U.K." *Management Accounting*, June 1995, pp. 42–44.

Booth, R. "Hitting the Target." *Management Accounting*, January 1995, p. 42.

Bromwich, M., and S. Inoue. *Management Practices and Cost Management Problems in Japanese-affiliated Companies in the United Kingdom*. The Chartered Institute of Management Accountants, U.K., 1994.

Cole, Robert E. "Learning from the Japanese: Prospects and Pitfalls." *Management Review*, September 1980, pp. 22–28 and 38–42.

Cooper, R., and C. A. Raiborn. "Finding the Missing Pieces in Japanese Cost Management Systems." *Advances in Management Accounting*, Vol. 4, 1995, pp. 87–102.

Daley, L., J. Jiambalvo, G. L. Sundem, and Y. Kondo. "Attitudes Toward Financial Control Systems in the United States and Japan." *Journal of International Business Studies*, Fall 1985, pp. 91–110.

Dutton, J. J., and M. Ferguson. "Target Costing at Texas Instruments." *Journal of Cost Management*, Fall 1996, pp. 33–38.

Freedman, J. M. "Target Costing." *Management Accounting*, April 1994, pp. 72–74.

Gagne, M. L., and R. Discenza. "Target Costing." *Journal of Business & Industrial Marketing*, Vol. 10, No.1, 1995, pp. 16–22.

Gietzmann, M., and S. Inoue. "The Adaption of Management Accounting Systems to Changing Market Conditions: Japanese Evidence." *British Journal of Management*, Vol. 2, 1991, pp. 51–55.

Howell, R. A., and M. Sakurai. "Management Accounting (and Other) Lessons from the Japanese." *Management Accounting*, December 1992, pp. 28–34.

Koons, F. J. "Applying ABC to Target Costs." *AACE Transactions*, 1994, pp. CSC.11.1–CSC.11.4.

Lee, J. Y. "Use Target Costing to Improve Your Bottom-Line." *The CPA Journal*, January 1994, pp. 68–70.

Martin, J. R., W. K. Schelb, R. K. Snyder, and J. S. Sparling. "Comparing U.S. and Japanese Companies: Implications for Management Accounting." *Journal of Cost Management*, pp. 6–14.

McMann, P., and A. J. Nanni, Jr. "Means Versus Ends: A Review of the Literature on Japanese Management Accounting." *Management Accounting Research*, Vol. 6, 1995, pp. 1–34.

Monden, Y. *Cost Management in the New Manufacturing Age*. Cambridge, MA: Productivity Press, 1992.

Monden, Y. *Toyota Management System: Linking the Seven Key Functional Areas*. Cambridge, MA: Productivity Press, 1992.

Monden, Y., and J. Lee. "How a Japanese Auto Maker Reduces Costs." *Management Accounting*, August 1993, pp. 22–26.

Ray, M. R. "Cost Management for Product Development." *Journal of Cost Management*, Spring 1995, pp. 52–60.

Sakurai, M. "The Influence of Factory Automation on Management Accounting Practices: A Study of Japanese Companies," in Kaplan, R. S.

(ed.), *Measures for Manufacturing Excellence*, Boston: Harvard Business School Press, 1990, pp. 39–62.

Scarbrough, P., A. J. Nanni, Jr, and M. Sakurai. "Japanese Management Accounting Practices and the Effects of Assembly and Process Automation." *Management Accounting Research*, Vol. 2, 1991, pp. 27–46.

Slipkowsky, J. N. "Is Japan the Key to Our Future?" *Management Accounting*, August 1993, pp. 27–30.

Tatikonda, L. U., and M. V. Tatikonda. "Tools for Cost-Effective Product Design and Development." *Production and Inventory Management Journal*, Second Quarter, 1994, pp. 22–28.

Worthy, F. S. "Japan's Small Secret Weapon." *Fortune*, August 12, 1991, pp. 48–51.

Yamamoto, T. "Toyota vs. the Rising Yen." *Tokyo Business*, July 1995, pp. 42–44.

Yoshikawa, T., F. Mitchell, and J. Moyes. *A Review of Japanese Management Accounting Literature and Bibliography*. The Chartered Institute of Management Accountants, U.K., 1994.

Yoshikawa, T., J. Innes, and F. Mitchell. "Cost Management through Functional Analysis." *Journal of Cost Management*, Spring 1989, pp. 14–19.

Yoshikawa, T., J. Innes, and F. Mitchell. "Cost Tables: A Foundation of Japanese Cost Management." *Journal of Cost Management*, Fall 1990, pp. 30–36.

Yoshikawa, T., J. Innes, and F. Mitchell. "A Japanese Case Study of Functional Cost Analysis." *Management Accounting Research*, Vol. 6, 1995, pp. 415–432.

Background Readings

Competition

Cusumano, M. A. *The Japanese Automobile Industry*. Cambridge: Council on East Asian Studies, Harvard University, 1985.

D'Aveni, R. *Hypercompetition*. New York: The Free Press, 1994.

Meyer, C. *Fast Cycle Time*. New York: The Free Press, 1993.

Porter, M. *Competitive Advantage of Nations*. New York: The Free Press, 1990.

Stalk, G. Jr., and T. M. Hout. *Competing Against Time.* New York: The Free Press, 1990.

Thomas, P. R. *Competitiveness Through Total Cycle Time.* New York: McGraw-Hill, 1990.

Williams, Jeffrey R. "How Sustainable Is Your Competitive Advantage?" *California Management Review,* Vol. 34, Spring 1992, p. 51.

Lean Production and the Lean Enterprise

Day, J. "The Power of Lean." *Chief Executive,* March 1995, pp. 50–51.

Fruin, W. M., and T. Nishiguchi. "Supplying the Toyota Production System" in Bruce Kogut (ed.), *Country Competitiveness: Technology and the Organizing of Work.* New York: Oxford University Press, 1993.

Harrison, B. *Lean and Mean: The Changing Landscape of Corporate Power in the Age of Flexibility.* New York: Basic Books, 1994.

Krafcik, J. F. "Triumph of the Lean Production System." *Sloan Management Review,* Fall 1988, pp. 41–52.

Morales, R. *Flexible Production.* New York: Polity Press, 1993.

Womack, J. P., and D. T. Jones. "From Lean Production to the Lean Enterprise." *Harvard Business Review,* March/April 1994, pp. 93–103.

Womack, J. P., and D. T. Jones. *Lean Thinking.* New York: HarperCollins, 1996.

Womack, J. P., D. T. Jones, and D. Roos. *The Machine That Changed the World.* New York: HarperCollins, 1990.

Product Development

Clark, K. B., and T. Fujimoto. *Product Development Performance.* Boston: Harvard Business School Press, 1991.

Clark, K. B., and T. Fujimoto. "Heavyweight Product Managers." *The McKinsey Quarterly,* No. 1, 1991, pp. 42–60.

Clark, K. B., and S. C. Wheelwright (eds.), *The Productivity Development Challenge.* Boston: Harvard Business School Press, 1994.

Imai, K., I. Nonaka, and H. Takeuchi. "Managing the New Product Development Process: How Japanese Companies Learn and Unlearn" in K. B. Clark et al. (eds.), *The Uneasy Alliance: Managing the Productivity-Technology Dilemma.* Boston: Harvard Business School Press, 1994.

Odagiri, H., and A. Goto. "The Japanese System of Innovation: Past, Present, and Future" in R. Nelson (ed.), *National Innovation Systems.* New York: Oxford University Press, 1993, pp. 76–113.

Technology Diffusion

Okimoto, D. *Between MITI and the Market.* Stanford: Stanford University Press, 1989.

Samuels, R. Jr. "Pathways of Technological Diffusion in Japan." *Sloan Management Review*, Spring 1994, pp. 21–34.

Quality

Garvin, D. A. *Managing Quality.* New York: The Free Press, 1988.

Crosby, P. B. *Quality Is Free.* New York: McGraw-Hill, 1979.

Hauser, P., and D. Clausing. "The House of Quality." *Harvard Business Review*, May-June 1988, pp. 63–73.

Ishikawa, K. *What Is Quality Control? The Japanese Way.* Englewood Cliffs, NJ: Prentice-Hall, 1985.

Index

ABC. *See* Activity-based costing
Acme Pencil Company, 197–220
Action champion, 232, 233
Activity-based costing (ABC), 54, 79, 232, 233, 355
Aiwa, 303, 307
Allowable cost, 87, 104–105, 108, 109–113, 114–116, 127, 355
Amoeba management system, 355
Analysis champion, 232, 233
As-if cost, 120, 355
Asahi Pentax, 282

Batch-and-queue, 355
Batch-and-queue product development, 27–29
Benchmark allowable cost, 113, 355
Benchmark costs, 105, 207
Best Topcon Program (TP), 322
Bunsha philosophy, 356

Camera, 35mm, 282–284
Canon, 282, 314, 315
Capacity utilization
   at Olympus, 293
Cardinal rule, 51, 78–79, 83, 108, 111–113, 116, 119, 120, 122–124, 124, 126, 127, 128, 147, 148, 162, 164, 215–216, 217–218, 219, 230, 356
Caterpillar, 37
Characteristics of the product, 174, 178–181, 194, 195
Checklist method
   at Isuzu, 136, 343
Chief engineer, 27, 51, 79, 146–147, 148–151, 152–153, 162, 163–164, 233, 257, 263, 264
   at Acme Pencil Co., 206
   at Toyota, 259
   role of, 142–144

Citizen Watch Company, xix, 97, 123, 125, 347, 349
Compact cameras, 283–284
Competitive analysis, 168
Competitive strategy, 30, 165
Complexity of the product, 174, 178–179, 195, 256
    at Komatsu, 191
Component-level target costing, 75, 139, 148, 166, 181, 182, 183, 185, 186–187, 188, 196
    at Acme Pencil Co., 216
    at Komatsu, 191
    at Nissan, 190
    at Olympus, 192
    at Sony, 192
    at Toyota, 190
Component-level target costs, 87, 153, 196
Computer simulations, 76, 91, 359
    at Nissan, 186, 245
Concept proposal stage, 132
Confrontation, 356
Confrontation strategy, xv, xxiii, 31, 36, 39, 42, 45, 47-48, 49, 59, 168, 169, 221
Consumer analysis, 81, 92, 94, 169, 171, 179, 242–246
    at Nissan, 171, 244
    at Olympus and Toyota, 171
Cost, defined, 32
    managing it, 38
Cost deployment flowchart, 132–133, 330–332, 343
Cost leader, 22, 30, 34, 40, 44, 46, 47, 169, 223, 356
Cost management, xx, xxi–xxiii, 37, 40–41, 47, 49–60, 61–63, 64, 72, 106, 130–131, 161, 221, 223, 224, 225, 228–234, 253, 319, 330, 352, 357
Cost management programs, xxii
Cost management techniques, 49, 53, 57, 59, 188, 232, 234

Cost reduction, xxi
    at Topcon, 321
Cost strategy map, 132, 332, 343
Cost tables, 266
Cost-down program, 356
Cost-reduction objective, 109
Cost-reduction objectives, 158, 231–233, 356
Cost-reduction targets, 131
Craft producers, 22
Current cost, 109, 111, 356
Customer analysis, 167, 173
    at Isuzu, 336
    at Komatsu, 191
Customer requirements, 167, 173, 175, 194
    at Komatsu and Olympus, 191
    at Toyota, 190
Customer satisfaction, 175

Daihatsu, 327
Degree of customer sophistication, 170–172
Degree of horizontal integration, 181–182, 185–186, 192, 227
Degree of innovation, 174, 177–178, 185, 187, 195
    at Komatsu, 191
    at Sony, 192
    at Toyota, 177
Design analysis
    at Komatsu, 275–276
Design engineers, 53
    at Toyota, 266
Design for manufacture and assembly (DFMA), 126–127, 356
Design teams, 110, 142, 164, 232
    role of, 151–152
DFM. See Design for manufacture and assembly
Differentiation, 356
Draft target cost, 116, 356
Drifting cost, 120–121, 127
Drifting target cost, 120–121, 356

Duration of product development, 174, 178, 180–181, 185, 187, 195, 196
  at Toyota, 180

Equipment investment at Toyota, 262
Expected cost, 157, 356
Expected selling price, 357

Feedback cost management techniques, 59
Feedforward cost management techniques, 59, 234
Final target cost, 357
First to market, 47, 202
First-look value engineering (VE), 133, 134–135, 138, 338, 357
First-mover advantages, 44–45, 168
Ford, 21, 358
Frequency of redesign, 174, 176–177, 185, 195
  at Komatsu, 191
  at Olympus, 176
Fuji Heavy Industries, 327
Functional analysis, 154, 177
  at Komatsu, 191, 276–278
Functional group management system, 357
Functionality, 165, 171, 175
  at Isuzu, 336
  at Olympus, 166
  defined, 32
  managing it, 37
Future requirements, 173, 185, 194

Gemini heater, 135
General Motors (GM), 340
GET method, 342
Group estimate by teardown (GET) at Isuzu, 342–343
Group leader, 58, 131, 293

Harness the entrepreneurial spirit, 38, 49, 57, 59
Higashimaru Shoyu Company, xix, 347
Hino, 327
Honda, 240, 327

Ibuka, Masaru, 301
Inamori, Kazuo, 349
Information sharing, xvii
Institute of Management Accountants, xvi, 61, 62
Intensity of competition, 167–169, 184, 186, 194, 196, 223, 234
Interorganizational cost management, xvii, 50, 52, 183, 228–234
Interorganizational costing, 139
IPS. See Isuzu Production System
Isuzu, 64, 129–130, 132–133, 135, 141, 327–345, 348
Isuzu Production System (IPS), 334

Japanese teardown method, 340
JIT. See Just-in-time
JKC, 348
Just-in-time (JIT), 334, 357

Kaizen, 124, 128, 357
Kaizen costing, xviii, 53, 56–57, 59, 132, 212, 357
Kamakura Iron Works, xix, 348
Keiretsu, 348, 357
Kenwood, 303
Kirin Brewery Company, xix, 348
Komatsu, 56, 65, 122, 140, 154–158, 269–279, 349
Kyocera Corporation, xx, 58, 349

Lean competitor, 46
Lean enterprise, xv, xix, 357
Lean enterprises, xvi, 46, 63, 113, 221

Lean leader, 44
Lean producer, 22, 29, 39
Lean product development, 28–29
Lean production, 21, 22, 227, 357
Lens shutters (LS) camera, 283–284
Life-cycle analysis, 77, 102
Life-cycle costing, 87
Life-cycle target costing, 180

Magnitude of up-front investment, 174, 178, 179–180, 185, 195
Market analysis, 51, 76, 121, 146, 167, 181, 186, 197, 223, 359
Market-driven costing, 74, 87, 166, 167–169, 185, 186, 187, 188, 196
  at Komatsu, 191
  at Olympus, 191–192
  at Sony, 192
  at Topcon, 193
  at Toyota, 190
Mass producer, xv, 358
MAST program at Topcon, 323
Matsushita, 303, 335
Maximize value approach, 336
Maximum allowable price, 33, 358
Maximum feasible values, 358
Mazda, 240, 327
Mini value engineering at Isuzu, 137, 339
Minimum allowable functionality, 34
Minimum allowable quality, 34
Minimum allowable value, 358
Minimum feasible price, 33
Minolta, 282
Mitsubishi Kasei Corporation, xx, 55, 327, 349
Miyota Company, xx, 349
Morita, Akio, 301, 305
Motoki, Masayoshi, 322

Multifunctional design teams, 48, 51, 233
Multifunctional teams, 25–29, 130

Nature of supplier relations, 181, 234
Nature of the customer, 167, 169–174, 194
New product models at Nissan, 241–251
Nidek, 315
Nihon Seiko, 159, 336
Nikon, 282, 315
Nippon Kayaku, 350
Nissan, 65, 91, 92, 93, 94, 100–101, 102–104, 116, 124, 130, 150, 158, 159, 165, 171, 172, 239–251, 327, 350
Nth-look approaches, 133–134
Number of products, 93, 106, 174, 174–176, 179, 185, 186, 195, 225, 315, 355

Olympus, 29, 37, 42, 65, 90, 93–94, 103–104, 116–119, 121, 125, 131, 281–298, 350
  product strategy, 165
Omachi, 350
Operational control, xviii, 53, 55–56, 59, 358

Perceived functionality, 31
Perceived quality, 31
Perfect waste-free cost, 114–115, 127, 358
Pilot project, 232–233
Planning stage, 132
Power over suppliers, 181, 183, 185, 192, 196
Price, 175
  defined, 32
Price point, 97

Procurement, 292
Product complexity, 178–179,
    185, 186, 196, 226–227
Product cost system at Nissan,
    250–251
Product costing, xviii, 53,
    59, 358
Product design, 225, 234
    at Toyota 255
Product designers, 87, 107,
    172, 233
    at Acme Pencil Co., 209–214
Product development
    at Komatsu, 270–274
    at Nissan, 246–247
    at Olympus, 191
    at Toyota, 177, 257–258
Product engineers, 199, 275
Product family, 27–29, 54
Product life cycle, 176
Product mix, 186
Product planning process
    at Komatsu, 270–271
Product strategy, 174,
    174–178, 194, 226–227
    at Sony, 192
Product-level target costing,
    75, 107–108, 127, 166,
    174–181, 185, 186, 188
    at Komatsu, 191
    at Nissan, 190
    at Olympus, 192
    at Sony, 192
    at Topcon, 193
    at Toyota, 190
Product-level target costing
    process, 119–126
Product-level target costs, 87,
    111–119
    at Acme Pencil Co., 208, 215,
    217–218
Production stage
    at Nissan, 249–250
Productivity analysis, 154, 157
    at Komatsu 158, 191,
    278–279
Profit management, 222–223

Profit maximizers, 41
Pseudo micro profit centers,
    xix, 57, 59

Quality, 165
    defined, 32
    managing it, 37
Quality function deployment
    (QFD), 126–127, 358

Rate of change of customer
    requirements, 172–173,
    185, 191, 194
Real micro profit centers, xix, 57,
    58, 59
Reengineering, 231
Research approach, 62
Reserve for production manager,
    148–149, 358
Responsibility centers, 55
Role of the accounting department,
    233

Sanno Institute of Management,
    335
Sanyo, 303
Satisfying customers, 223–224
Sato, Yoshihiko, 329, 340, 345
Second-look value engineering,
    358
    at Isuzu, 133, 134, 135, 138,
    338–339
SEI (Sumitomo Electric Industries),
    351
Self-guided teams, 129, 231–233
Setting component-level target
    costs, 150–151
Setting retail prices and sales targets
    at Toyota, 258–264
Setting selling prices, 166
Setting target costs of major
    functions, 140
Setting the product-level target cost,
    108
Sharp, 303

Shionogi, 351
Single lens reflex (SLR) camera,
 283–284
Single-piece flow, 358
Single-piece flow philosophies, 29
Site selection, 62–64
SLR camera, 284
Sony, 64, 65–66, 91, 100, 123,
 125, 165, 299–312, 351
Sony Walkman, 126, 173,
 300–303
Sophistication of customers,
 184, 194
SRS. *See* Symbiosis research system
Stamina line, 304, 309
Standard cost, 124, 128
Strategic cost-reduction challenge,
 110–119, 127, 358
Structured product line, 92
Sumitomo Electric Industries. *See*
 SEI
Supplier creativity, 159
Supplier negotiations, 161
Supplier relations, xvii, 53, 87,
 110, 159, 169, 183–184,
 185, 186, 227, 247
 at Acme Pencil Co., 216,
  217–218
 at Isuzu, 336–337
 at Komatsu, 191, 273–274
 at Nissan, 190
 at Olympus, 192
 at Topcon, 193
 at Toyota, 190, 255–256
Supplier selection, 160, 164
 at Isuzu, 161
 at Komatsu, 162
 at Nissan, 161
Supplier-base objectives, 159
Supplier-base strategy, 181–182,
 192, 194
Suppliers, 177, 187
Supply chain, xvii
Survival triplet, xv, 31, 36, 38,
 47–48, 165, 171, 221,
 358

Survival zone, xvi, 32, 44, 88,
 167, 185, 358
Sustainable competitive advantage,
 43, 359
Sustainable cost advantage, 30
Suzuki, 327
Symbiosis research system (SRS)
 at Isuzu, 335

Taiyo Group, 58, 351
Target cost system at Olympus,
 289–292
Target cost-reduction objective,
 109–113, 119–122, 126, 359
Target cost-setting process,
 144–151
Target costing, xvii, xxii, xxiii,
 29, 50, 59, 71–80, 87,
 166–167, 169, 172, 173,
 175, 176, 178, 188, 221,
 230–233, 359
Target costing benefits, 180, 183,
 185, 194, 195, 196
Target costing process, 165, 168,
 184, 186, 188, 194, 219
 at Komatsu, 190–191, 274–279
 at Nissan, 188
 at Olympus, 191–192
 at Sony, 192, 196
 at Topcon, 193
 at Toyota, 188–190, 257
Target costing system
 at Nissan, 165, 241–251
 at Topcon, 196
 at Toyota, 165, 257
Target costing systems
 subprocesses of, 166
Target profit margin, 88,
 100–101, 359
Target selling price, 94, 165,
 177, 359
Teardown, 133, 168
 approaches, 135–136
 at Isuzu, 340–342
 cost teardown, 341

dynamic teardown, 340
group estimate by teardown
(GET), 342–343
material teardown, 341
matrix teardown, 341
process teardown, 341
static teardown, 341
unit-kilogram price method, 342
Teardown analysis, 359
Temporary competitive advantage,
359
TMW (Tokyo Motor Works), 352
Topcon, 43, 55, 66, 90, 96–99,
130, 165, 313–325, 352
Toshiba, 303, 317, 322
Total quality management, 359
TOV. *See* Turn-out-value system
Toyo Valve, 159, 336, 352
Toyota, 66, 90, 95–99,
121–122, 140, 142–144,
151–152, 165, 171, 188,
240, 253–267, 327, 353
TQM programs, 37–38
Trial production stage
at Komatsu, 272–273
Tsuchiya, Hiroshi, 335
Turn-out-value system
at Topcon, 317–322

Unavoidable waste-free cost,
114–115, 127, 359
Understanding future requirements,
173–174, 194
Up-front investment, 106, 176,

178, 179, 185, 191, 192,
226

Value analysis (VA), 131–132
at Topcon, 323–325
Value engineering, xxii, 29, 51,
78, 80–85, 108, 110, 128,
129, 132, 136, 221,
230–233, 244–246, 359
at Isuzu, 133–137, 335–340
at Toyota, 264–265
Value engineering (VE), 126–127
Value engineering reliability
program at Isuzu, 339–340
Value target (VT) at Isuzu,
335–337
Variance analysis, 55
at Topcon, 321
VE. *See* Value engineering

Waste-free cost, 113–114

Yamanouchi, 353
Yamatake-Honeywell, 353
Yokohama, 354
Yuasa, 159, 336

Zero defects (ZD), 359
at Topcon, 322
Zero-look value engineering
at Isuzu, 133–134, 138,
337
Zero-look VE, 359

BOOKS FROM PRODUCTIVITY PRESS

Productivity Press publishes books that empower individuals and companies to achieve excellence in quality, productivity, and the creative involvement of all employees. Through steadfast efforts to support the vision and strategy of continuous improvement, Productivity Press delivers today's leading-edge tools and techniques gathered directly from industry leaders around the world. *Call toll-free (800) 394-6868 for our free catalog.*

Companywide Audit Pack
Strategic Direction Publishers (ed.)

Any improvement effort first requires a thoughtful assessment of the current state of your organization. This powerful package, compiled from the work of many leaders in the field, is on state of the art company auditing. The portfolio is designed for business managers creating business strategies. Contained in this single comprehensive portfolio are the details of 12 different self-assessment audits. Whether you have done company audits in the past, or now recognize how critical audits are to your strategic plans, this portfolio will provide you with many new and exciting ideas, a deeper understanding, the steps for the audit process, and most importantly, how to use the data you collect to create a powerful and effective company strategy.
ISBN 3-908131-00-6 / 872 pages / $295.00 / Order BMAF-B277

Cost Management in the New Manufacturing Age
Innovations in the Japanese Automotive Industry
Yasuhiro Monden

Up to now, no single book has explained the new cost management techniques being implemented in one of the most advanced manufacturing industries in the world. Yasuhiro Monden has taught the principles of JIT in the United States and now brings us firsthand insights into the future of cost management based on direct surveys, interviews, and in-depth case studies available nowhere else.
ISBN 0-915299-90-9 / 198 pages / $30.00 / Order COSTMG-B277

Cost Reduction Systems
Target Costing and Kaizen Costing
Yasuhiro Monden

Yasuhiro Monden provides a solid framework for implementing two powerful cost reduction systems that have revolutionized Japanese manufacturing management: target costing and kaizen costing. Target costing is a cross-functional system used during the development and design stage for new products. Kaizen costing focuses on cost reduction activities for existing products throughout their life cycles, drawing on approaches such as value analysis. Used together, target costing and kaizen costing form a complete cost reduction system that can be applied from the product's conception to the end of its life cycle. These methods are applicable to both discrete manufacturing and process industries.
ISBN 1-56327-068-4 / 400 pages / $50.00 / Order CRS-B277

Handbook for Productivity Measurement and Improvement
William F. Christopher and Carl G. Thor, eds.

An unparalleled resource! In over 100 chapters, nearly 80 frontrunners in the quality movement reveal the evolving theory and specific practices of world class organizations. Spanning a wide variety of industries and business sectors, they discuss quality and productivity in manufacturing, service industries, profit centers, administration, nonprofit and government institutions, health care and education. Contributors include Robert C. Camp, Peter F. Drucker, Jay W. Forrester, Joseph M. Juran, Robert S. Kaplan, John W. Kendrick, Yasuhiro Monden, and Lester C. Thurow. Comprehensive in scope and organized for easy reference, this compendium belongs in

every company and academic institution concerned with business and industrial viability.
ISBN 1-56327-007-2 / 1344 pages / $90.00 / Order HPM-B277

Integrated Cost Management
A Companywide Prescription for Higher Profits and Lower Costs
Michiharu Sakurai

To survive and grow, leading-edge companies around the world recognize the need for new management accounting systems suited for today's advanced manufacturing technology. Accountants must become interdisciplinary to cope with increasing cross-functionality, flexibility, and responsiveness. This book provides an analysis of current best practices in management accounting in the United States and Japan. It covers critical issues and specific methods related to factory automation and computer integrated manufacturing (CIM), including target costing, overhead management, Activity-Based Management (ABM), and the cost management of software development. Sakurai's brilliant analysis lays the foundation for a more sophisticated understanding of the true value that management accounting holds in every aspect of your company.
ISBN 1-56327-054-4 / 352 pages / $50.00 / Order ICM-B277

Making the Numbers Count
The Management Accountant as Change Agent on the World Class Team
Brian H. Maskell

Traditional accounting systems hold back improvement strategies and process innovation. Maskell's timely book addresses the growing phenomenon confronting managers in continuous improvement environments. It unmasks the shortcomings of the management accountant's traditional role and shows the inadequacy of running a business based on financial reports. According to Maskell, in a world class organization, the management accountant can and should take the lead in establishing performance measures that make a difference. Empowering frontline workers and middle managers to effectively improve their operations is one way to do it.
ISBN: 156327070-6 / 150 pages, illustrations / $29.00 / Order MNC-B277

To Order: Write, phone, or fax Productivity Press, Dept. BK, P.O. Box 13390, Portland, OR 97213-0390, phone 1-800-394-6868, fax 1-800-394-6286. Outside the U.S. phone (503) 235-0600; fax (503) 235-0909. Send check or charge to your credit card (American Express, Visa, MasterCard accepted).

U.S. Orders: Add $5 shipping for first book, $2 each additional for UPS surface delivery. Add $5 for each AV program containing 1 or 2 tapes; add $12 for each AV program containing 3 or more tapes. We offer attractive quantity discounts for bulk purchases of individual titles; call for more information.

Order by E-Mail: Order 24 hours a day from anywhere in the world. Use either address:
To order: **service@ppress.com**
To view the online catalog and/or to order: **http://www.ppress.com/**

Quantity Discounts: For information on quantity discounts, please contact our sales department.

International Orders: Write, phone, or fax for quote and indicate shipping method desired. For international callers, telephone number is 503-235-0600 and fax number is 503-235-0909. Prepayment in U.S. dollars must accompany your order (checks must be drawn on U.S. banks). When quote is returned with payment, your order will be shipped promptly by the method requested.

Note: Prices are in U.S. dollars and are subject to change without notice.

OTHER PUBLICATIONS FROM IMA'S FOUNDATION FOR APPLIED RESEARCH, INC.

The IMA Foundation for Applied Research, Inc. (FAR) is the research affiliate of the Institute of Management Accountants. The mission of the Foundation is to develop and disseminate timely management accounting research findings that can be applied to current and emerging business issues.

Management Accounting in Support of Manufacturing Excellence
Profiles of Award-Winning Organizations
Richard L. Jenson, James W. Brackner, and Clifford R. Skousen
200 pages   96311/$40

The researchers studied eight organizations that received the Shingo Prize for Excellence in Manufacturing to find out how management accounting has adapted to the new manufacturing environment of increased global competition, more flexible and efficient manufacturing processes, and more demanding customers. The research findings provide a variety of approaches used to streamline the management accounting system.

Management Accounting Issues in Cellular Manufacturing and Focused-Factory Systems
Dileep G. Dhavale
268 pages   96310/$60

This book is written for financial managers and management accountants who have installed or are planning to install cellular manufacturing and a focused-factory system. It provides background material, fundamentals, and methods for developing a system to deal with these manufacturing concepts. Information in the book is based on current industry practices, as ascertained from an in-depth survey and practice-oriented written material.

Measuring Corporate Environmental Performance
Best Practices for Costing and Managing an Effective Environmental Strategy
Marc J. Epstein
160 pages   95301/$40

The author surveyed the management of a number of large organizations to discover how environmental costs and strategies affect decision making.

He describes the best practices of corporate environmental performance and the different methods companies can use to reach the ultimate goal of concurrent engineering. The book provides specific steps on how to integrate environmental considerations into all design and management decisions.

The Theory of Constraints and Its Implications for Management Accounting

Eric Noreen, Debra Smith, and James T. Mackey
160 pages   94300/$40

The theory of constraints provides a coherent and focused management framework within which management accountants can ply their trade. It has profound implications for the way cost and financial reporting systems and management controls are constructed. The authors visited sites that had put the theory of constraints into practice. The book contains detailed case studies of seven of these companies.

An ABC Manager's Primer

Straight Talk on Activity-Based Costing
Gary Cokins, Alan Stratton, and Jack Helbling
65 pages   93282/$20

The primer, written in simple, clear language and extensively illustrated, gives a head start for action in implementing an ABC program. Issued in collaboration with CAM-I.

Activity-Based Management in Action

Patrick L. Romano
160 pages   94289/$40

Twenty-five thought-provoking articles cover the evolution from activity-based costing (ABC) to activity-based management (ABM). The articles, first published in *Management Accounting,* are grouped in seven sections with introductions and provide a clear perspective of the ways ABC/ABM has been used and adapted.

Implementing Activity-Based Cost Management

Moving from Analysis to Action
R. Cooper, R. S. Kaplan, L. S. Maisel, E. Morrissey, and R. M. Oehm
336 pages   92268/$40

This research study documents key issues and results of implementing activity-based cost management, based on real-life examples and case studies of eight companies.

STATEMENTS ON MANAGEMENT ACCOUNTING

Statements on Management Accounting (SMAs) present the views of IMA regarding management accounting issues. In their development, the Statements are subjected to a rigorous exposure process and are published only upon approval by IMA's Management Accounting Committee.

Individual SMAs $7.50

Complete set of all SMAs $60

1A   Definition of Management Accounting

1B   Objectives of Management Accounting

1C   Standards of Ethical Conduct for Management Accountants

1D   The Common Body of Knowledge for Management Accountants

1E   Education for Careers in Management Accounting

2A   Management Accounting Glossary

4A   Cost of Capital

4B   Allocation of Service and Administrative Costs

4C   Definition and Measurement of Direct Labor Cost

4D   Measuring Entity Performance

4E   Definition and Measurement of Direct Material Cost

4F   Allocation of Information Systems Costs

4G   Accounting for Indirect Production Costs

4H   Uses of the Cost of Capital

4I   Cost Management for Freight Transportation

4J   Accounting for Property, Plant, and Equipment

4K   Cost Management for Warehousing

4L   Control of Property, Plant, and Equipment

4M   Understanding Financial Instruments

4N   Management of Working Capital: Cash Resources

4O   The Accounting Classification of Real Estate Occupancy Costs

4P   Cost Management for Logistics

4Q   Use and Control of Financial Instruments by Multinational Companies

4R   Managing Quality Improvements

4S   Internal Accounting and Classification of Risk Management Costs

4T   Implementing Activity-Based Costing

4U   Developing Comprehensive Performance Indicators

4V   Effective Benchmarking

4W   Implementing Corporate Environmental Strategies

4X   Value Chain Analysis for Assessing Competitive Advantage

4Y	Measuring the Cost of Capacity
4Z	Tools and Techniques of Environmental Accounting for Business Decisions
4AA	Measuring and Managing Shareholder Value Creation
5A	Evaluating Controllership Effectiveness
5B	Fundamentals of Reporting Information to Managers
5C	Managing Cross-Functional Teams
5D	Developing Comprehensive Competitive Intelligence
5E	Redesigning the Finance Function

To order: Contact Customer Orders & Information Department, Institute of Management Accountants, 10 Paragon Dr., Montvale, NJ 07645-1760; phone (201) 573-9000 x 278 or (800) 638-4427 x 278; fax (201)573-9507. Send check payable to Institute of Management Accountants or charge to your credit card (American Express, Visa, MasterCard accepted). All orders must be prepaid.

Order by e-mail: jpirard@imanet.org. To view the online catalog or learn more about IMA and its services: http://www.imanet.org.

IMA members in good standing are entitled to a 20% discount. To receive this discount, you must furnish your IMA member account number. Faculty members of colleges and universities, nonprofit libraries, and college and retail bookstores are entitled to a 40% discount. Educational discounts and member discounts may not be combined.

Shipping: All orders within the continental United States are shipped by UPS or first-class mail. Minimum shipping and handling charge per shipment is $6 for 1–3 items; $9 for 4–10 items; $12 for 11–20 items; and $20 for 21 or more items. All other orders are sent via air printed matter at a minimum of $40 or 40% of the retail price of the books, whichever is higher.

Note: Prices are in U. S. dollars and are subject to change without notice.